HOW TO INCREASE CAPACITY OF YOUR WAREHOUSE

TIME TESTED TECHNIQUES TO HELP YOU FIND SPACE YOU DIDN'T KNOW YOU HAD.

Art Liebeskind

Howard Way & Associates

Published By

Industrial Data & Information Inc.

http://www.IDII.com

HOW TO INCREASE THE CAPACITY OF YOUR WAREHOUSE

Time tested techniques to help you find space you didn't know you had.

Art Liebeskind

ISBN 978-0-9800734-4-7
Printed in USA. History November 2010 - First Edition

Industrial Data & Information, Inc. (IDII)
2448 East 81st Street – Suite 2044
Tulsa, OK 74137
U.S.A.
918-292-8785 Voice
918-512-4132 Fax
www.IDII.com orders@idii.com

Table of Contents

ABOUT THE AUTHOR

Art Liebeskind has been a warehousing consultant for more than 35 years. As president of Howard Way and associates, he has taught warehousing seminars and completed projects in the United States, Canada, England, Germany, Italy, Austria, Sweden, Japan, Australia, Thailand, Malaysia, Indonesia, Singapore, The Phillipines, China, Hong Kong, South Africa, Egypt, Tunisia, Mali, Mexico, Chile, Argentina, Colombia, Peru, and Venezuela.

Mr. Liebeskind has designed and implemented six fully automated warehouses and a large number of more conventional facilities.

He published and wrote the Howard Way Letter, a guide to productivity in the warehouse (The archives of which are available on a CD with a search engine). He is the creator of WAREHOUSE MASTER WINDOWS, a cost-effective warehouse management system for personal computers.

He holds a bachelor's degree in mechanical engineering and a master's degree in industrial engineering from Cornell University.

These tips have been collected from his extensive personal experience. He will do his best to answer emails with questions/comments.

Contact the Author

Art Liebeskind art@howardway.com

ACKNOWLEDGEMENTS

This book was prepared by the staff of Howard Way and associates. The drawings were prepared using Corel Draw and Sketchup graphics software. We acknowledge the contributions and guidance of our publisher, Philip Obal of Industrial Data & Information Inc.

The following sources gave permission to include photographs and drawings:

- The Raymond Corporation
- Frazier Industries

This book too is dedicated to *Patricia Gordon Liebeskind* in continued appreciation of her help, encouragement and boundless support.

Chapter 1: INTRODUCTION

This is the second book in our warehousing series. It is titled "How To Increase The Capacity of Your Warehouse" that span "Time Tested Techniques to help you find space you didn't know you had."

This book will analyze various Equipment, Packaging and Space Management Systems and how they influence space utilization. General "Tips" will be presented in these chapters. Specific "Tips" or recommendations will be presented for each functional area of the warehouse. For each functional area, note the specific methods to increase the utilization of space and realize that they may apply to other areas as well. We will reiterate Tips where necessary to underline the fact that they may apply to several situations

We hope that this book will serve as a spark plug to jolt you into an organized search for hidden space in your warehouse. We will point you to functional areas in the warehouse operation in which specific topics will be pursued that will lead to increased utilization of space.

In every seminar that I lead, I ask a question the answer to which I can almost unfailingly predict. That question is, *"How many of you, are running out of space?"* The response is an almost unanimous raising of hands. We are all running out of space in our present facility or are ready to acquire a new facility (green field, purchase, or rented). If not, we are destined to suffer with stopgap methods of storage that either foster double handling or cause us to "lose inventory hidden behind other items".

THE BASIC STEPS TO IMPROVE SPACE UTILIZATION ARE TO:

I. Define the facility
II. Define the stored material
III. Utilize the CUBE

The best way to start thinking about finding space that you didn't know you had is to make yourself aware of the details of your facility and the properties of that which you store. The best way to do this is a formal approach shown below.

Tip No. 1: DEFINE THE FACILITY

Develop hard data about:
Area (Square Meter, Square Ft)
Clear height to the structure (Meters, feet)
Floor strength or loading limits (kg/M2, lb./Ft2)
Bay size (Distance between columns in both directions)
Multi-floor or one floor?
Rental or ownership cost (money per square unit or overall cost per year)

Tip No. 2: DEFINE THE STORED MATERIAL

List the types of material and their properties:

- Value
- Fragility
- Security
- Flammability

List The Packaging Arrangements:

- Pallets
- Cases
- Drums
- Bulk

Tip No. 3: CONSIDER OTHER REQUIREMENTS:

- Lot number control,
- Expiration requirements, and
- Need for stock rotation.
- Material type or category
- Relative unit and estimated total value of each category
- Stackability of material for each category
- Packaging arrangements for each category
- Flammability for each category
- Lot control requirements for each category
- Rotation requirements for each category
- Security requirements for each category
- Picking criteria for each category (unit picks, case picks, pallet quantity picks or mixtures of the foregoing)

The formal collection of the information mentioned above will in many cases suggest solutions that were not apparent from more casual observations. In addition, careful considerations of certain limitations caused by the needs will often save you from expensive mistakes.

Tip No. 4: UTILIZE THE CUBE

Simply stated, material is stored in three dimensions. A consideration of area alone is not realistic. In general, most solutions involve higher storage in the same area and/or narrower aisles. Other solutions involve denser storage and the utilization of space that is taken but is empty. An example is the storage of a short pallet in a tall rack slot. The slot is "occupied" but the full potential is not used. That situation is called **HONEYCOMB** and many ways to reduce it will be suggested.

THE BASIC TOOLS AND CONCEPTS

I. Handling Equipment/Aisle width/Storage Height

II. Storage Equipment/Storage Height

III. Packaging/Palletization

IV. Management Control Systems

Refer back to the tools and concepts outlined in each of these sections when implementing the suggestions made for each specific area of the warehouse.

Tip No. 5: ESTABLISH A ROUTINE FOR MAKING SPACE RATIONAL DECISIONS

Use the checklist following to make sure you have considered all of the factors that influence a change in layout or procedure. Evaluate each factor and assess its priority in making the decisions.

The check sheet shown covers many factors that influence the layout and therefore influence the best use of space. This check sheet can be used both to plan a new warehouse and to improve an existing warehouse. Think hard about each item and figure out for yourself just how important it is in your operation. Formalizing the routine goes a long way towards considering all aspects.

THE BASIC APPROACH TO A RATIONAL LAYOUT

A Check Sheet

FACILITY DEFINITION

Area ____sq.ft Height ________ft below steel
Floor strength__lbs/sq.ft. Scale drawings Yes/No?
Bay Size: __ft __inches X __ft __inches
Annual Cost per sq. Ft. $_______ Rent OR OWN
NumberTruck Doors ____- Number Rail Doors ___

MATERIAL DEFINITION

Define each family or category	Yes/No	☐
Total value for each category	Yes/No	☐
Relative unit value for each category	Yes/No	☐
Stackability For each category	Yes/No	☐
Stability for each category	Yes/No	☐
Packaging, (each/case/pallet)	Yes/No	☐
Flammability by category	Yes/No	☐
Rotation requirements	Yes/No	☐
Lot Control	Yes/No	☐
Pick Criteria (each,case,pallet)	Yes/No	☐
Security evaluation	Yes/No	☐

OUTSIDE INFLUENCES

- Where does material come from? ☐
- Where does material go? ☐
- Who is your customer? ☐
- Is the warehouse a profit center? ☐
- What are the warehouse goals? ☐

SETTING PRIORITIES (Which are critical?)

- Distribution Costs ☐
- Customer Service ☐
- Minimal Inventory(High turns) ☐
- Maximum Space Utilization ☐
- Throughput ☐
- Accuracy of Orders ☐

Formalize the collection of the above information. Write it down and interview affected areas of your organization for input into the warehouse needs required. SPEAK TO YOUR CUSTOMERS. Find out what they need and want. Design accordingly!

Tip No. 6: KNOW THE VALUE OF SAVING SPACE

Any space or square footage saved will not disappear from the facility, and unless the space is either used by the warehouse to store additional goods or transferred to another department, it will **not** constitute a saving. If however, additional warehouse space is needed, it may be acceptable to value the space made available at a rate comparable to rental space in an outside warehouse or even at the avoided cost of new construction.

Conversations with the cost department will turn up the fact that the warehouse is charged a per square foot rate for the space it occupies. This rate may also be used as an equivalent for space saved. Remember that space made idle or empty is not considered a savings unless it is put to productive use. The use of racks and narrow aisle handling equipment may increase the usable cube in a warehouse. If this avoids outside warehousing, by all means count not only the out-of-pocket cost of the public warehousing but also the saving in trucking cost by avoiding the transport to the warehouse.

In the case of a proposed new building, where it is possible to calculate the space required per load stored, using the present storage configuration and a proposed narrow aisle high cube system, count the space saved at the new-construction cost estimate for a square foot of building. The costs of necessary racks and handling equipment would be considered part of the capital cost of the proposed alternative.

THE WAREHOUSE AREAS TO BE CONSIDERED ARE:

1. Receiving
2. Storage
3. Order Picking
4. Packing Area / Order Staging Area
5. Shipping Area
6. Locations

Remember please that many “tips” apply to more than one department or area in the warehouse. Make it a routine that when a tip is implemented, consider its value in all other parts of the operation as well.

Chapter 2: HANDLING EQUIPMENT, AISLES, STORAGE HEIGHT

The choice of available material handling equipment is very wide and each offers the capability of using a particular aisle width and height capability.

Tip No. 7: USE NARROWER AISLES
Changing from a wide aisle to a narrow aisle can significantly increase the storage density in your warehouse.

Tip No. 8: STORE HIGHER
Similarly, storing taller can do the same by utilizing unused height in your building. Either or both may be accomplished with the equipment types shown following. You can be conservative and still gain huge amounts of space. Remember to justify your change by expressing the value of the added space and comparing that to the cost. The value may come from cancelling third party storage, avoidance of building an expansion or even avoiding the cost of building a whole new warehouse.

Fork trucks have changed over the years with the effect of ever narrowing the aisle needed to handle pallets. Around time of World War II, the counterbalanced fork truck itself was invented and suddenly the warehouse was freed of the necessity of hand stacking boxes or cases of merchandise eight feet high (the limit for a tall and strong man) or even twelve feet high with two men working as a hand up and stack team. Suddenly we had the ability to utilize heights that were unattainable before. This meant that in a given area we could store more material in the same area. Even then this represented a trade-off since the aisle requirements of 12 to 14 feet were greater than the narrower aisle needed for manual storage. Fortunately the trade-off was one that favored the increased utilization of the space.

Today, the counterbalanced fork truck is still the standard of the industry even though many proven alternatives exist today. In the early 1950s some new types of truck appeared. These new devices were the narrow aisle straddle truck and the reach or extend truck. The straddle truck has a number of weaknesses, the most important of which is a lack of flexibility in the pallet sizes and types that it can handle. This is because of the necessity for the pallet base to nestle between the two outriggers of the truck. In addition it was restricted to handling wing pallets (those with a base smaller than the deck)

The reach or extend truck overcame the difficulties with a slight sacrifice in aisle width by adding a mechanism to extend the pallet beyond the outriggers and set it directly on the floor or rack. It was now possible to utilize narrow aisle equipment with a larger diversity of pallet sizes.

Another logical development was the deep reach truck. This truck has an increased extension capability so that the the forks could reach into a double deep pallet rack. By storing two loads deep, the overall requirements for load area are reduced. This in turn makes the area for loads stored even smaller. Remember that no improvement comes without cost. The down side with a double reach pallet truck comes when the lot sizes to be stored are not two or more pallets. In other words you may want the rear pallet before you need the pallet in the front. This of course increases the handling effort.

In the 1960s, automatic storage and retrieval systems (sometimes called Stacker cranes) were introduced and these utilized captivesmachines going up and down each aisle to bring the desired pallet to the load and unloading station at the head of the aisle. These were larger structures that ran as tall as 90 feet or more. These are most suited to large operations with extremely high throughput of material.

A compromise called the turret truck or swing reach truck operates by turning the mast in the aisle rather than the whole truck. Because of this development, it can put away or retrieve a standard GMA pallet (48 by 40) in an aisle as small as five feet wide. In addition, this truck can stack loads as high as 45 feet. This combination of aisle width and stacking height can yield an area per load of three square feet . Keep firmly in mind the fact that the cost of building the building higher is relatively less per foot man the cost of adding additional square area. This fact means building a taller structure has an extreme economic advantage.

Shown below are the general types of equipment available and the heights and minimum aisle widths attainable with each. Remember that the minimum aisle may be attainable only with a great deal of maneuvering of the truck. In some cases, the better choice will be a bit more than the absolute minimum attainable. As always in the logistics field, choices are often a compromise and a little bit of "sub-optimization."

In the examples that follow, assume an 85 foot by 85 foot piece of the warehouse. In the example of a **counterbalanced truck,** the aisles are 12 feet wide and the height is 3 pallets high. For the **narrow aisle reach truck** assume an 8 foot aisle (one could go narrower but this is an easy in and out width that allows for good access). The height could now go as high as 5 pallets high(if the clear height allows). The **walkie Reach truck** will probably only stack 3 high. The turret truck example assumes a 5+ foot aisle and a potential height utilization of 8 pallets high. The maximum capacity is shown with each layout.

AISLES & HEIGHTS FOR PALLETS ON RACK

Following are some lift truck options that allow rack aisles and storage heights to be improved.

COUNTERBALANCED FORK TRUCK

AISLE WIDTH APPROXIMATELY 12 FEET
STORAGE LEVELS = 2 to 4 levels
Be aware of the collapsed height of the mast so that the truck will pass through existing door ways. The operator sits aboard the truck

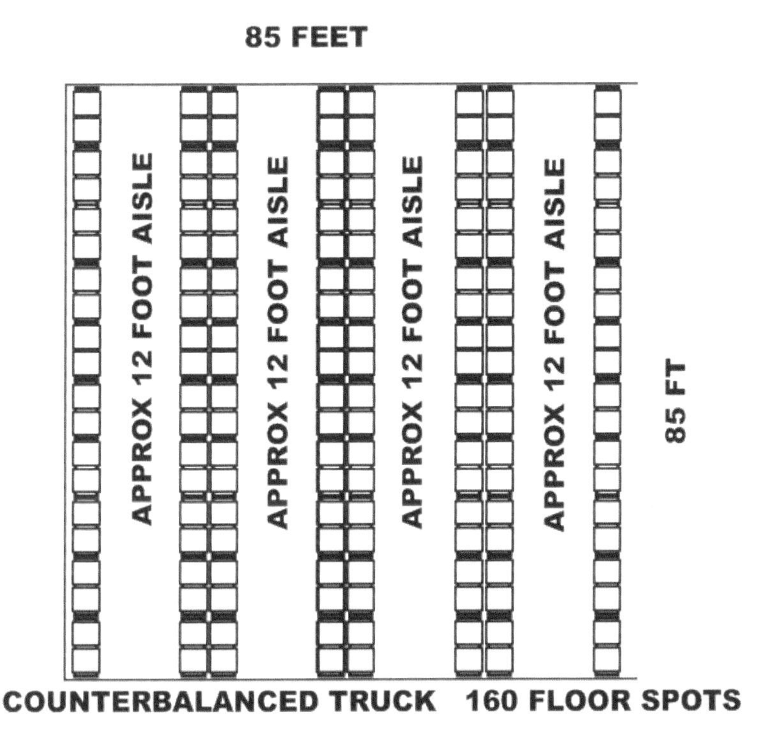

Capacity = 160 spots x 3 high = 480 pallets

The counterbalanced fork truck is the most commonly used truck. The aisles are wide and the vertical lift is somewhat limited. It is probably what you are presently using and a relayout for a narrower aisle and higher lift truck will offer additional storage in the same space. An advantage of the counterbalanced truck is that it serves dual purpose in its ability to go in and out of trailer trucks for loading and unloading.

WALKING REACH STACKER TRUCK

AISLE WIDTH APPROXIMATELY 6 FEET
STORAGE LEVELS = 2-4
The Operator walks behind the truck

In a small warehouse, a riding truck is rarely needed. In fact, mounting and dismounting can take longer than the travel time. The model shown above is a walkie reach truck which can handle a wide range of pallet widths. Similar trucks that straddle the pallet can also be used with a specific and standard pallet

RIDING REACH TRUCK

AISLE WIDTH APPROXIMATELY 7 FEET
STORAGE LEVELS = 2-4
The Operator sits sideways in the cab

In a larger warehouse, a riding truck is appropriate. The one illustrated above has an operator seated at right angles to the aisle. Other types utilize a standing operator. These models generally can offer a higher lift than the walkie trucks.

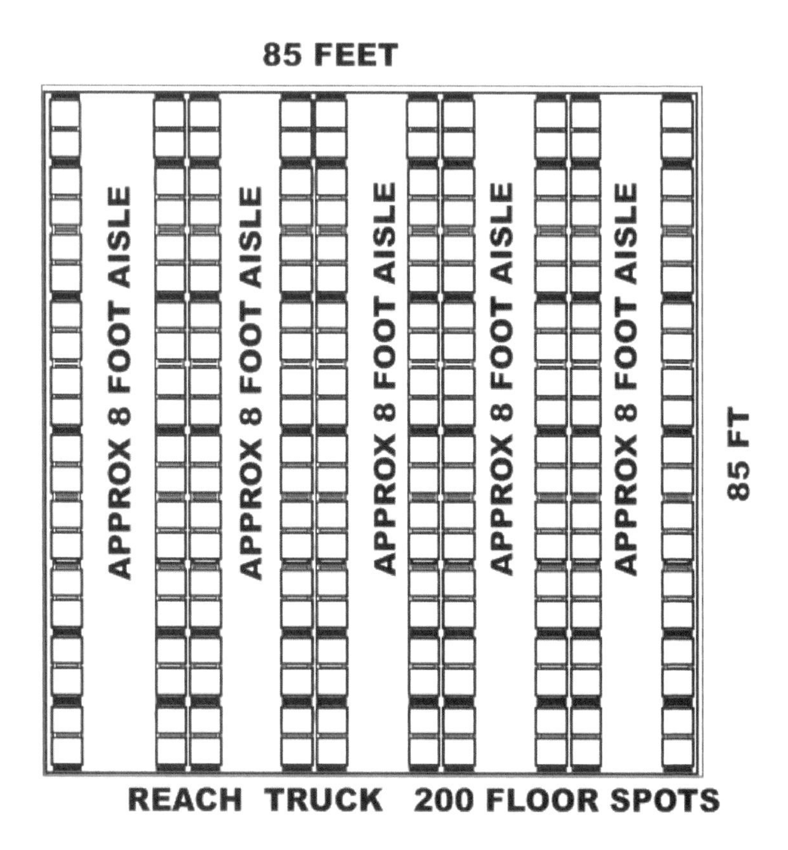

Rider version capcity = 200 spots x 5 high = 1000 pallets
Walkie version capacity = 200 spots x 3 high = 600 pallets

DEEP REACH TRUCK

AISLE WIDTH APPROXIMATELY 8-9 FEET
STORAGE LEVELS = 2-5

Assume Stacking 5 Pallets High
Similar to reach truck ,stores pallets two deep. Requires slightly wider aisle than reach truck

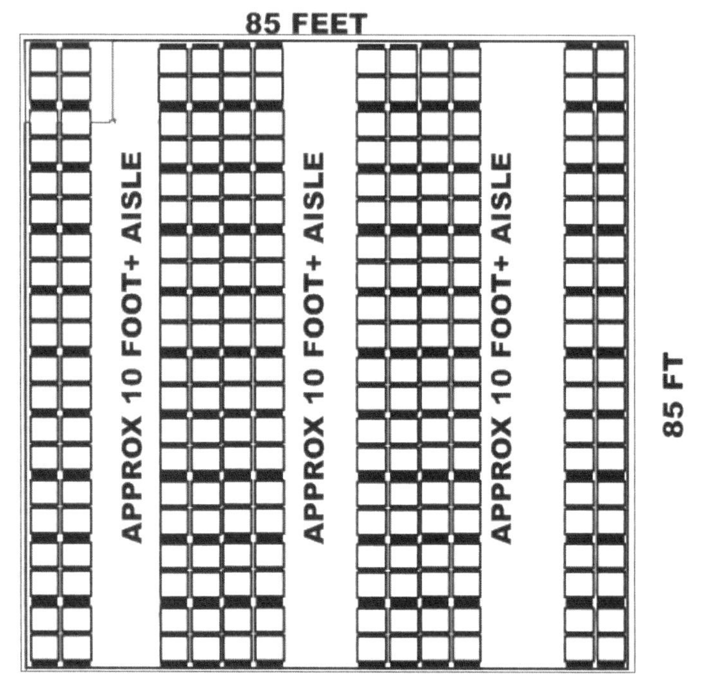

DOUBLE DEEP REACH TRUCK 240 FLOOR SPOTS
Capacity = 240 spots x 5 high = 1200 pallets

Tip No. 9: **STORE DOUBLE DEEP**

Recognize that this is best if you have many identical pallets so that if they are stored two deep, you may choose either the front pallet or the rear. If not too many pallets of one particular SKU exist in the warehouse, you may find that you need to remove the front pallet, set it down, retrieve the rear pallet and then replace the one set aside.

TURRET TRUCK

AISLE WIDTH APPROXIMATELY 5 FEET
MAXIMUM LIFT = 42 Feet

The Turret Truck can use an aisle as narrow as 5 feet plus (Depending on pallet size). The lift height can exceed 40 feet. Models are available where the operator goes up with the load (better control and placement). Other models leave the operator below and can be equipped with height and position indicators to simplify put away and withdrawal. This is a practical solution where there are many unique SKUs since only one pallet deep is stored.

Higher reaching trucks must have a very level floor to run on. A special level floor specification is needed and that can add $1.00 or so to the cost per square foot.

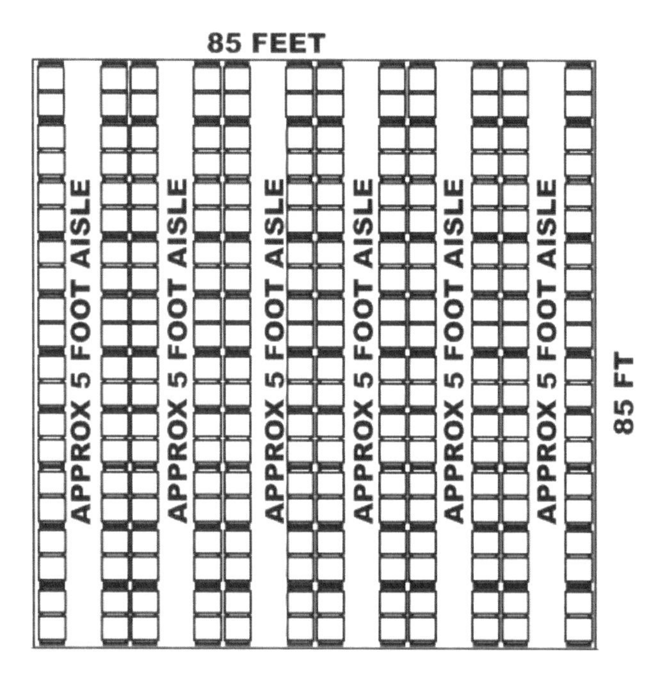

TURRET TRUCK 240 FLOOR SPOTS

Capacity Calculation:

8 loads high= 240 spots x 8 high = 1920 pallets

A Summary of the advantages of narrower aisles and taller storage can be seen in the table following.

Comparison Of Space Gains From Narrower Aisles And Greater Height

Equipment	Footprint S/F	Number Of Tiers	Area Per Load In S/F	Number Of Loads In 10,000 S/F
Counterbalance forklift CB	40 S/F	3 Tiers	13.3 S/F/LD	752
Narrow Aisle Reach NA	26.7 S/F	5 Tiers	5.3 S/F/LD	752
Turret Truck Swing VNA	22.5 S/F	7 Tiers	3.2 S/F/LD	752
Storage/Retrieval Truck SRT	20.8 S/F	11 Tiers	1.9 S/F/LD	752

AISLES AND HEIGHTS FOR SMALL GOODS AND CASES ON PICK SHELVES

Following are some options for utilizing the cube and going higher in the order picking areas. Most small parts or "eaches" picking locations are limited to shelves about 8 feet high. Another common method is to pick from the lower sections of pallet rack. Both methods are limited to the vertical reach of the order selector standing in the aisle.

Tip No.10: USE TALLER SHELVES

Going higher to utilize the cube can be accomplished in a number of ways from very simple to major changes.involvig vehicles or mezzanines.
Following are some options that allow greater height to be utilized. It should be mentioned that aisles should already be at a relatively narrow width to allow the selector to pick from both sides with minimum travel back and forth.

Tip No. 11: USE PICK CARTS WITH LADDERS

This is an inexpensive way to allow order pickers to have a higher reach. Taller shelves will better utilize the height of the room (UTILIZE THE CUBE)

Tip No. 12: **USE A SMALL PARTS PICK MACHINE**

Small parts and cases can be accessed with a relatively new device called a “man up” pick cart. This electrically powered vehicle travels in extremely narrow aisles and is extremely maneuverable. It is generally restricted to a height of about 20 feet.

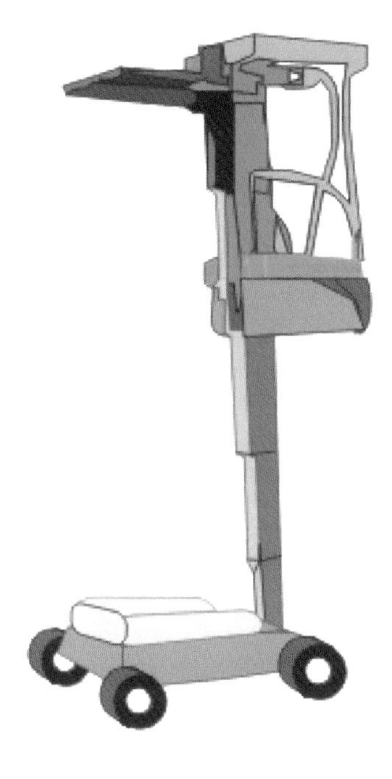

This equipment is particularly suited to picking of small parts from shelves in the 20 foot tall range. The operator goes up and selects the goods then deposits them on a carrier at the front of the vehicle. It is often helpful to install side guides in the aisle so that steering is unnecessary.

Tip No. 13: **CONSIDER ORDER PICKER TRUCKS**

The ability to access tall heights in a relativelynarrow aisle is one of the attributes of the ORDER PICKER TRUCK. The operator goes up with the forks. This vehicle can carry a pallet for picking and an operator to the pick face. It allows utilization of all of the available warehouse height for reachable pick spots. Order Picker Trucks can go to a height of 30 Feet

ORDER PICKER TRUCK

Tip No. 14: BUILD A STRUCTURAL MEZZANINE

Another approach is to use a Mezzanine or artificial "second floor" built above the existing shelf pick area. This can be a structural floor that is free standing above the existing shelves (See below) or an actual extension of the shelves with drop in floor plates (See illustration next page). Both styles require built in steps for personnel access. In addition, large quantities of material in and out of the upper levels dictate a "landing spot" for pallet loads drop and pickup or a conveyor or elevator type load lifter. Picked merchandise can also be brought down to the shipping level by use of an inexpensive gravity conveyor

Tip No. 15: USE THE 80/20 RULE

Put the fastest moving material below in the lower level. This will reduce the time spent climbing to the upper level.

STRUCTURAL MEZZANINE
Courtesy Porta-King Building Systems

MEZZANINE FORMED BY STACKED SHELF SECTIONS

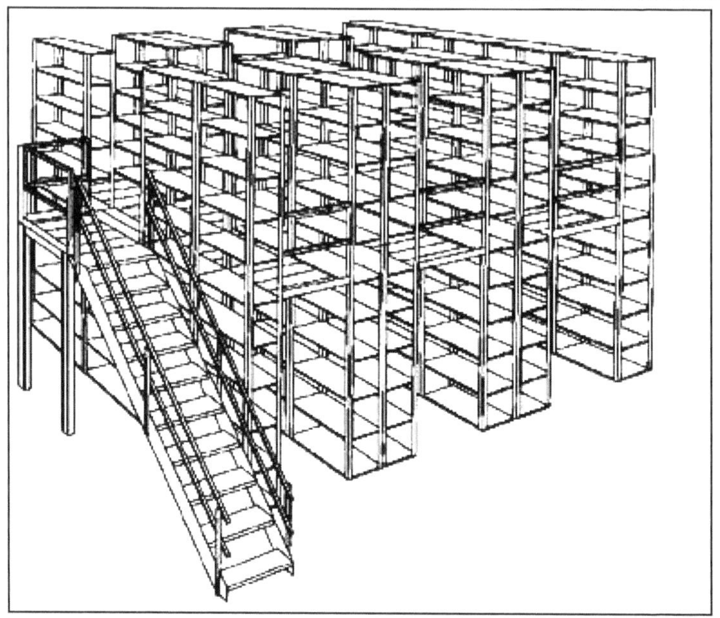

SHELF SUPPORTED MEZZANINE

Tip No. 16: **BUILD A SHELF SUPPORTED MEZZANINE**

An efficient and flexible way to build a mezzanine is to stack shelf sections (stressed of course for the planned load) one atop the other. Expanded metal flooring is dropped in place and other accessories are added as needed. A beneficial point is that this method of utilization of height is considered fiscally and tax wise as equipment and as such is subject to a fast write off. The equivalent addition of building floor space would be "real estate" and would be subject to a long depreciation period similar to a building.

Another benefit of this type of mezzanine is that if it is decided later, to utilize tall pick vehicles, the floor can be removed and the result is a very tall set of shelves that can be serviced without climbing stairs.

Chapter 3: STORAGE EQUIPMENT, STORAGE HEIGHT

There are many types of storage equipment available to facilitate taller or more dense storage of inventory. The most basic concept is storage of material directly on the floor without even the use of a pallet.(See Below) Of course this limits the height attainable and also makes any equipment aided handling difficult.

Tip No. 17: STORE GOODS ON A PALLET

The next step up from loose floor storage is the storage of goods on a single height pallet resting on the floor.(See next Page) A little easier to handle perhaps but still not utilizing the air space above. The basic thought here is that from a storage viewpoint, the higher that one stores above a given "footprint on the ground"*, the more efficient is that square area utilized This is really the basis or "Utilize The Cube", a cliché because it is so very true.

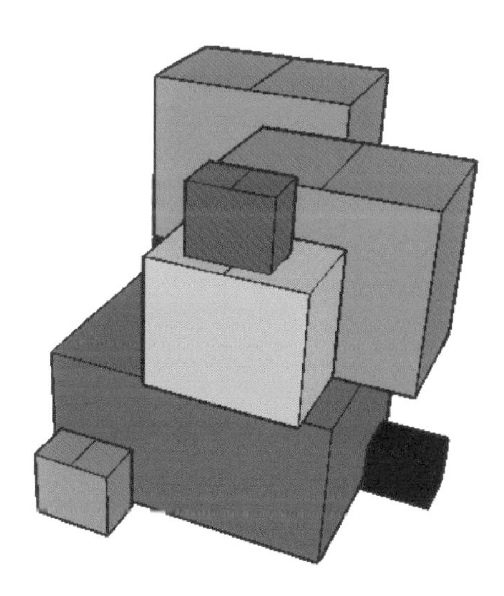

STORAGE ON THE FLOOR

* The area of the pallet plus ½ of the aisle serving it

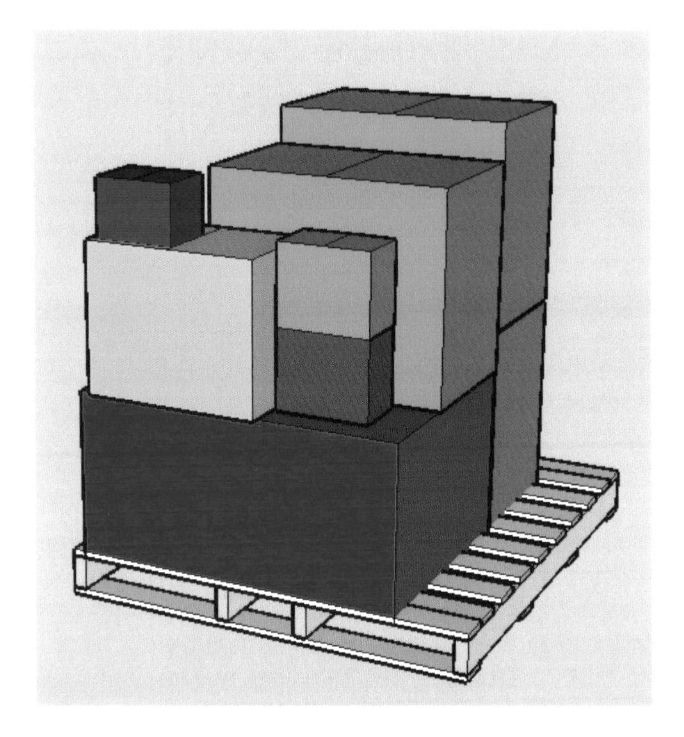

SAME MATERIAL ON A PALLET
The use of the pallet makes the goods easier to retrieve or move. This alone is an advantage.

Tip No. 18: STACK PALLETS ON FLOOR

On the following page, we see two pallets stacked one on the other. If the loads are stable and no crushing of the bottom load is anticipated, we have just doubled the storage in the same area or footprint. The accessability may suffer somewhat in a mixed pallet such as the one shown but with solid pallets of the same goods, accessability is not as much of a problem. Depending on the strength of the load and the stability of the pattern, it is common to see even more tiers of stacked pallets.

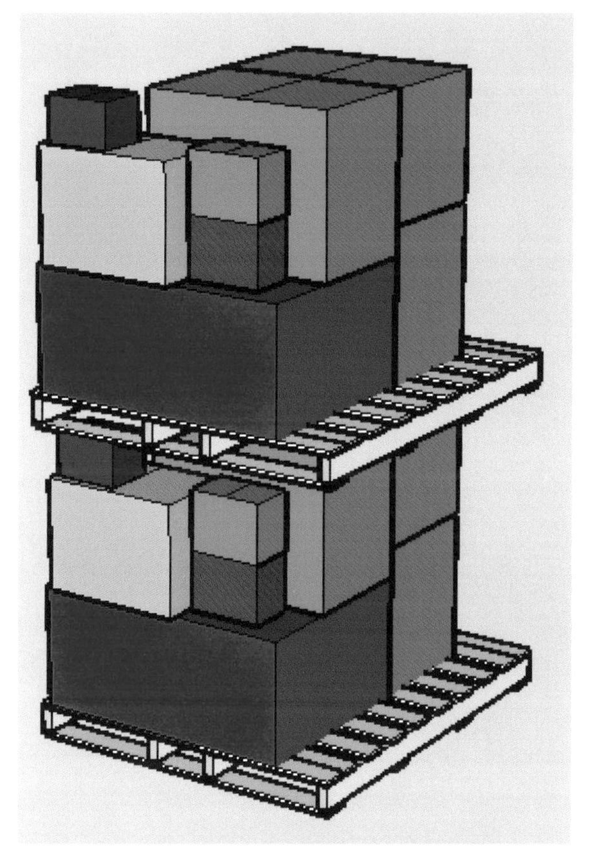

TWO PALLETS STACKED TWO HIGH

Tip No. 19: USE STACKING FRAMES

A very flexible piece of storage equipment, The Pallet Stacking Frame is shown on the following page. This allows material to be stacked higher without weight bearing on the material below. Pallet Stacking frames are similar to rack except that to move one unit, all above must be handled or lifted off. Depending on weight, several units may be lifted at one time.

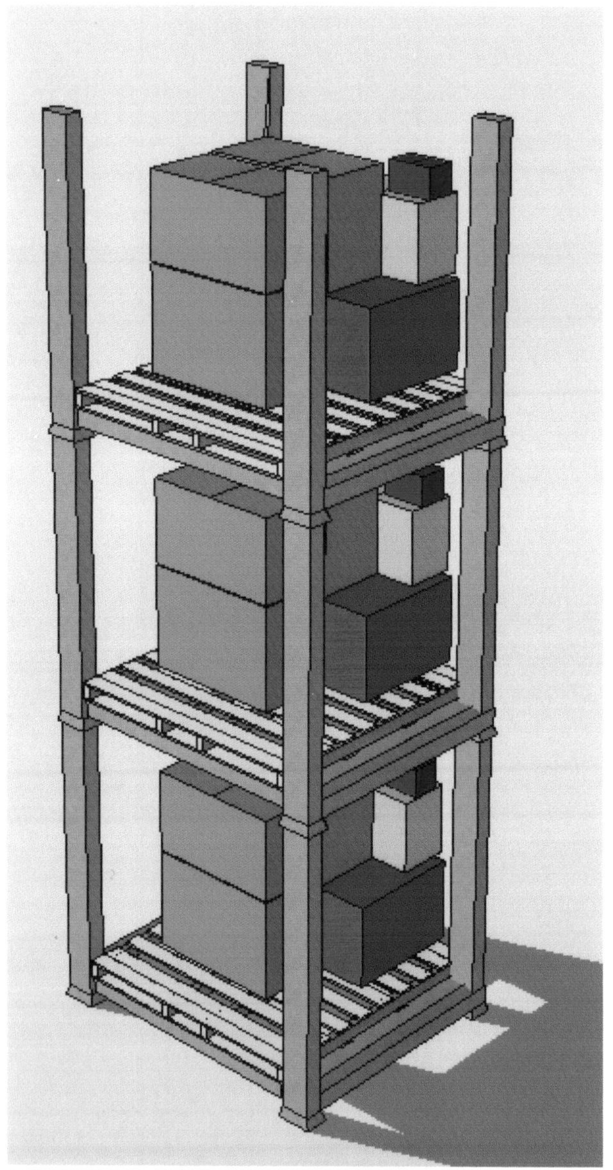

PALLET STACKING FRAMES

SIMPLE SELECTIVE PALLET RACK

Tip No. 20: **CONSIDER SELLECTIVE PALLET RACK**

Another piece of equipment that is useful to save space is the pallet rack as shown above. Now, we can store as high as the combination of truck lift height, rack height and a consideration of sprinkler or fire insurance considerations allow. Note that we have tripled the storage that we had with a single pallet high. We have "utilized the cube"

There are many types of racks and some of them are shown in the following pages.

SELECTIVE PALLET RACK

The selective pallet rack is the most used rack in industry. It is versatile and relatively inexpensive. Selective rack may be constructed of rolled steel or made up of structural steel sections.

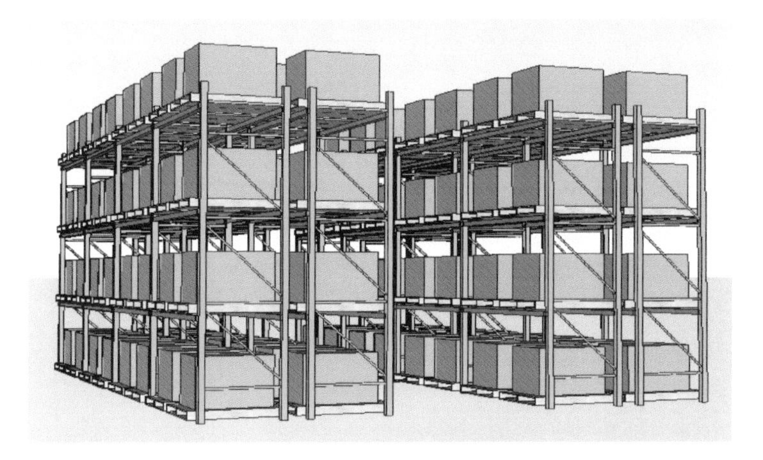

TYPICAL SELECTIVE PALLET RACK CONFIGURATION

Note that the typical arrangement is two racks on each side of the aisle. The layout usually finishes with a single rack if there is a wall or an important activity aisle not related to the racks.

CANTILEVER RACK

Tip No. 21: CANTILEVER RACK FOR LONG GOODS

Cantilever rack like the above are well suited for long or irregular storage. The arms can carry plywood or metal decks which make the racks excellent for furniture of varying lengths.

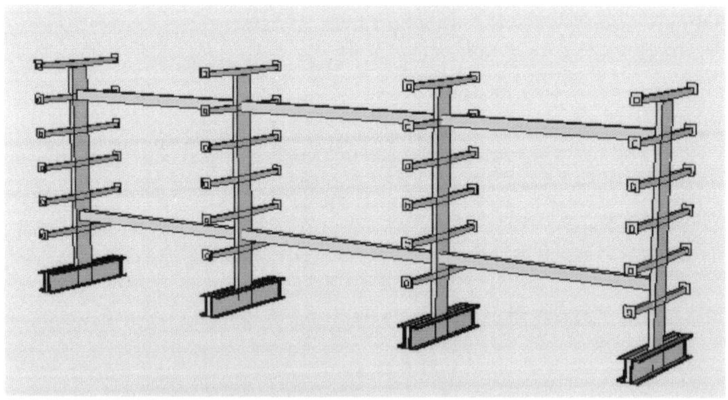

CANTILEVER PIPE RACK

Note that this rack has capacity for storage on both sides of the central upright. This type of rack is particularly suited to the storage of tubes, pipes and long narrow items.

Tip No. 22:EMERGENCY STORAGE IN A CONTAINER

If you are really short of space, a used shipping container can provide covered and protected storage outside the shelter of the warehouse. It may not be efficient storage but sometimes it can be the solution to a temporary peak.

Tip No. 23: USE "PUSHBACK"RACKS FOR DEEP STORAGE AND FEWER AISLES

Pushback racks are a solution to the need for increased density of storage. A number of pallets are loaded from the single aisle. The second in, pushes the first back into the "rack track" and so on to capacity. It is last in, first out which is not always acceptable. If the material is fungible, *(Todays vocabulary word: all are the same and any load is just as suitable as any other.)* the system can vastly increase warehouse capacity and greatly reduce the area used for aisles. The illustration on the next page shows in more detail how the racks are built

PUSHBACK RACKS 4 DEEP Courtesy Frazier Industrial

.

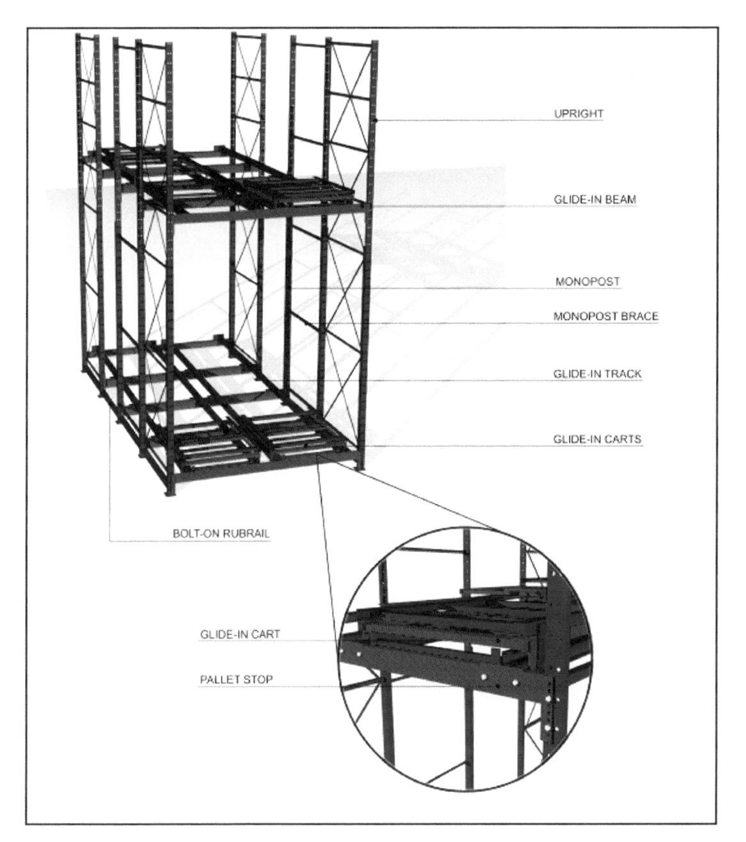

PUSH BACK RACK DETAILS Courtesy Frazier Industrial

Tip No. 24: ACHIEVE DEEP STORAGE WITH DRIVE-IN OR DRIVE-THROUGH RACKS

Drive in and drive through racks are really a refinement of deep storage (Deep Storage will be discussed in detail later) with the ability to go higher without crushing the pallets below. As such, it is subject to the disadvantages of deep storage. Frankly, both types often exhibit a large amont of "honeycomb" or empty spaces. Also, a drive in rack favors last in-first out. On the other hand, the drive through rack while it accomodates first in-first out, takes up 2 aisles as seen in the following illustration for drive through racks..

DRIVE IN RACK

See below for an explanation of how both work.

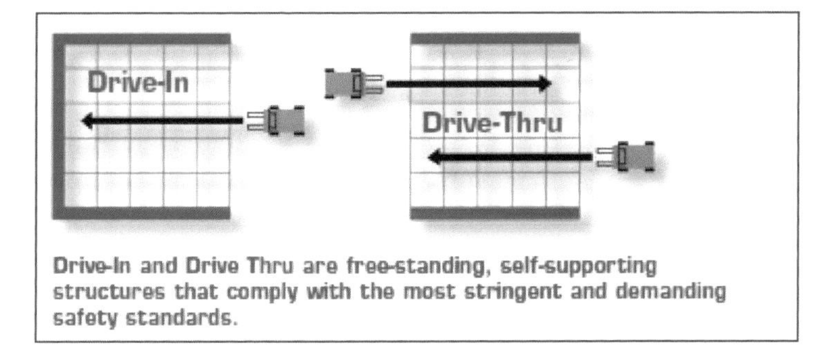

Drive-In and Drive Thru are free-standing, self-supporting structures that comply with the most stringent and demanding safety standards.

DRIVE IN AND DRIVE THROUGH RACKS

Tip No. 25: USE STACKING BINS FOR LOOSE GOODS

Stacking bins are similar to stacking frames with the added advantage of being able to hold loose or randomly placed items just like a big basket. An added advantage is that they easily fold and store as may be seen on the next page. They also can be had with a drop door that allows access to the contents even when the bin is part of a stack.

STACKING BIN

Like the pallet frames, multiple bins can be lifted and moved at one time. This becomes an advantage when an inside bin of the stack is the one that is needed.

STACKED WIRE BINS

FOLDED AND STORED STACKING BINS

As seen above, 10 folded bins are easily stored in one floor position. Stacking bins are often used in the food industry because being steel or stainless steel, they may be sanitized with steam or some other method of cleaning.

Tip No. 26: **CONSIDER MOVING AISLE RACKS**

MOVEABLE PALLET RACKS

The moveable aisle pallet system is an elegant but costly way to save space. The individual racks slide on a track so that in effect, the aisle location is moving about the various pairs of shelves. The sections are often individually powered by gearmotors to facilitate their movement. In the example above, five aisles that would be needed for a conventional layout are reduced to one "Moveable aisle".

The other part of the ever present trade-off is that while space is conserved, accessability is reduced. There is a lot of movement necessary when a pallet is needed from inside the cluster. Many sections must be moved for access.to a particular pallet in the closed stack.

SHELVING

Shelves are most often the solution for small quantities and sizes. They come in a variety of constructions and vary in cost as they become more sophisticated.

Tip No. 27: USE EASILY ADJUSTABLE SHELVES

In all types of shelving, it is important to acquire a type that is easily adjustable as to shelf height placement. Ordinary nut and bolt adjustments are costly to erect and difficult to change. Rarely will they be adjusted to fit changing sizes of goods stored.

SIMPLE OPEN SHELVING
The simplest type is open shelving. Although it can be acquired in its simplest form as shown above, many accessories are available. Particle board shelves are shown above but steel shelves or plywood shelves are available as well.

CLOSED SHELVING

Closed shelving which can be upgraded from open shelving with the addition of standard panels, provides more protection to the material stored. Many accessories are available such as dividers, bin type shelves, hangers and a variety of storage enhancements. **Please remember that shelves are essentially a small version of racks. Therefore, the same or similar space conservation principles apply to shelving as well as racks.**

Tip No. 28: USE A MOVABLE AISLE SHELF SYSTEM

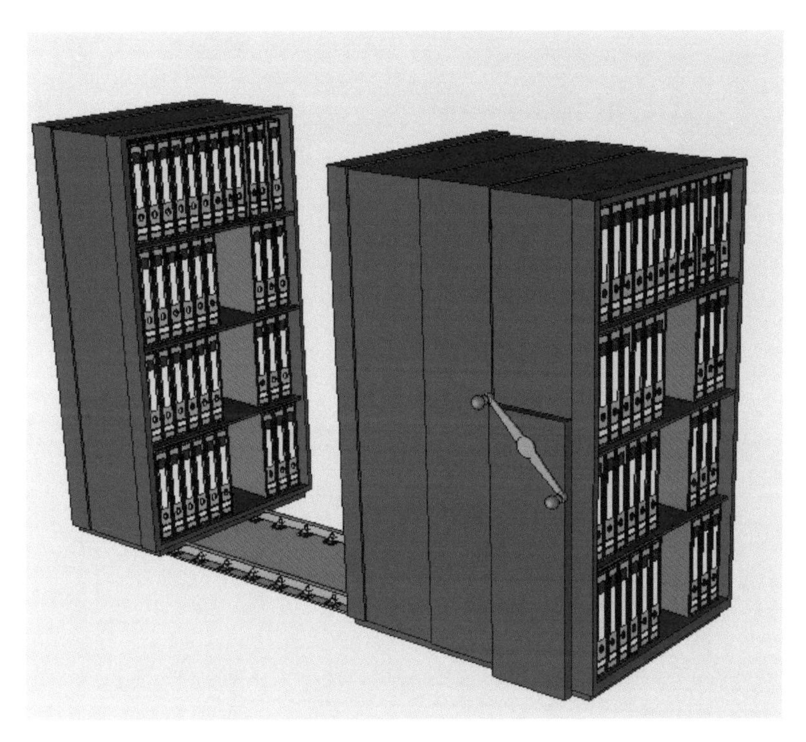

The moveable aisle shelf system is also an elegant but costly way to save space. The individual units slide on a track so that in effect, the aisle location is moving about the various pairs of shelves. The other part of the ever present trade-off is that while space is conserved, accessability is reduced. There is a lot of sliding necessary to reach an item in the closed stack. A similar concept is available for pallet racks as mentioned earlier but is generally much too pricy for it's capacity. A similar trade off is evident in this case when an item is needed from inside the cluster, many sections must be moved for access.

Tip No. 29: **USE A MODULAR DRAWER SYSTEM**

A special breed of drawer system has been gaining favor for the storage of small parts under tight control. This system has very heavy-duty drawers mounted on high-quality roller bearings that permit parts storage weights of up to 400 pounds per drawer. Heights of drawer systems vary from counter height to eye-level height and, under certain conditions, may be multiple stacked for semi-high-rise storage to be serviced by a stockpicker truck. Many styles of dividers, drawer boxes and tool holders are available with which to subdivide the drawers. They are extremely effective at conserving space when used for small parts or tools. Since the cost is fairly high, the space saved must be evaluated as part of the justification for this equipment. A flexible locator or addressing system is a must for rapid stock withdrawal.

Tip No.30: STACK THE DRAWERS HIGH

HIGH RISE MODULAR DRAWER SYSTEM

This is a case of not only “utilizing the cube” but also taking advantage of the organized storage within the drawers. Excellent for small parts and tooling. Note that the picking vehicle carries the operator up and in reach of the various drawers. An accurate locator system is really a necessity to make a system like this work efficiently.

Tip No. 31: **USE A HORIZONTAL CAROUSEL**

The horizontal carousel saves space because multiple units can be placed very close together with little or no aisle in between. Carousels bring material to the order picker rather than the order picker going to the goods. Carousels are a seductive solution and must be carefully evaluated. They should NOT hold fast movers but rather a wide selection of fairly slow movers. Fast-moving merchandise would need to be replenished so often that the advantage of reduced travel would be lost.

HORIZONTAL CAROUSELS Courtesy Diamond Phoenix

Tip No. 32: USE A VERTICAL CAROUSEL

The vertical carousel goes "round and round" but in a vertical plane. This uses the cube wisely and saves considerable space In addition, it offers considerable security since the small access door can be closed and locked. Similar cautions apply with the vertical c arousel as with the horizontal carousel. It is best for slow and small parts rather than very fast movers. In order to avoid excessive replenishment, the inventory of a fast moving item would have to be quite large and take up a lot of real estate in this fairly costly piece of equipment.

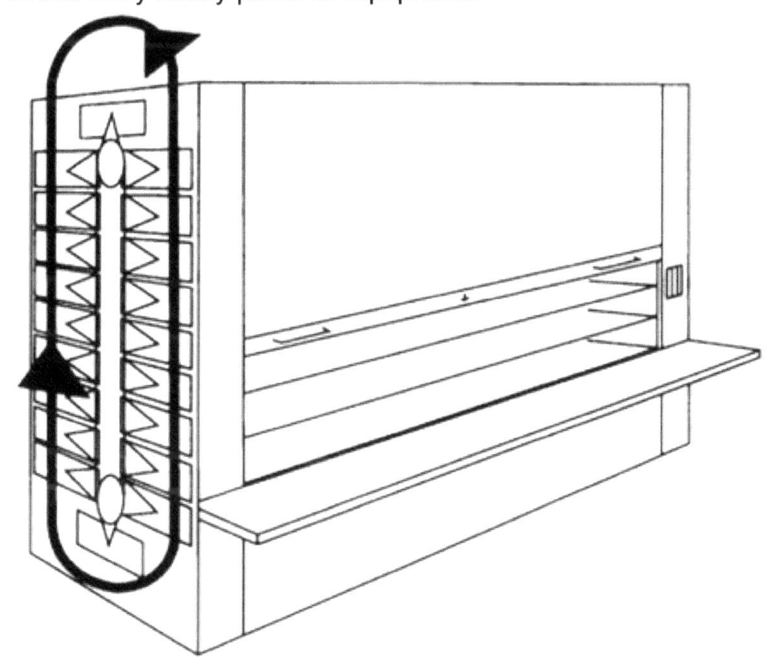

VERTICAL CAROUSELS Courtesy Diamond Phoenix

Tip No. 33: VERTICALLY STACK CAROUSELS

Both horizontal and vertical carousels can be stacked and therefore utilize great height as in mezzanines or tall storage machines or the equivalent of an order picker truck.

Tip No. 34: BUILD TALLER WITH NARROWER AISLES

Fact: Building costs rise more slowly per cubic meter in a taller building, and we store in cube, therefore taller buildings are more cost efficient. Equipment is written down more rapidly than buildings and so the cash flow is better when the major expenditure is on equipment. A rack supported building is considered equipment rather than real estate and will depreciate faster for tax purposes.

A graphic example of how this works may be seen in the following example of real costs and layouts. The area per load is calculated, the building costs are estimated (by building contractors) and the rack cost per load are all calculated for a range of configurations of height and aisle widths.

The building designs are real, The heights and dimensions are from actual layouts. The costs per square area are from builders. The rack costs are from suppliers
Although these costs were calculated some time ago, in a place not necessarily yours, they can be used today at your location. The reason is that we are looking at “differences", not absolutes.

- The costs of all options have risen proportionally.
- The costs for different states, provinces or countries are also proportional
- The **difference** per load stored remains quite accurate.

ASSUMPTIONS

- **Load Details: 48" x 40" load, 48" high including pallet to fit with lift clearance in a 53" opening. Weight = 2700 pounds. Estimates based on a 2,000 load system.**

- **Rack Details: Shelf beams 92" long. Two loads per opening. Frame depth of 42 inches. 12 bays per row. Structural steel rack based on specs from American Institute of Steel Construction.**

- **Building Details: No site work. Spread footing foundations. 5" reinforced concrete slab. Prefab steel building. Air rotation heating to 50 degrees. No plumbing and normal hazard sprinkler system. 20 foot candles of light, 400 ampere service. No doors or truck docks.**

Although the costs and the physical details may vary with time or location, the space results hold true and the costs are usable by comparing the differences between sets of options.The charts following give an immediate and dramatic cost demonstration of the advantages of tall, narrow aisle structures over short conventional aisle facilities.

THE SPACE CHART

The SPACECHART is derived from an elementary consideration of the actual floor space required to store a load of goods in the warehouse. (See next page ,"THE AREA IS THE SAME") This shows an analysis of two of the simplest warehouse configurations: single loads stored on a 12-foot aisle, and double-tiered loads stored on a 12-foot aisle. If we assume a 48" x 40" load with side clearances 4 inches on each side, the load area is 41 x 41, without consideration of the aisle. Since loads are usually stored on either side of the aisle, we must assign one half of the aisle to a single load. One half of the aisle (6 feet) plus a load depth of 4 feet equals 10 feet. Multiply this by the load width of 4 feet and find an area of 40 square feet. As inefficient as this sounds, there are many loads stored in just this way throughout today's warehouses.

The second part of the figure below is the same, except that a second load has been stacked on top of the first. The 40-square-foot "footprint" is identical, but now it is holding two loads. Dividing the 40-foot footprint by 2 yields a space per load of 20 square feet. This is the reasoning used to build the SPACE CHART. We have assumed a load typical of many found in our warehouses, and the clearances are typical of those recommended by many rack manufacturers On the following page, you will see the “SPACE CHART” . This useful tool is derived from an extension of the simple illustration above using real dimensions and methods to give a range of “Areas per Load” for a wide variety of aisle widths, storage heights and equipment choices.

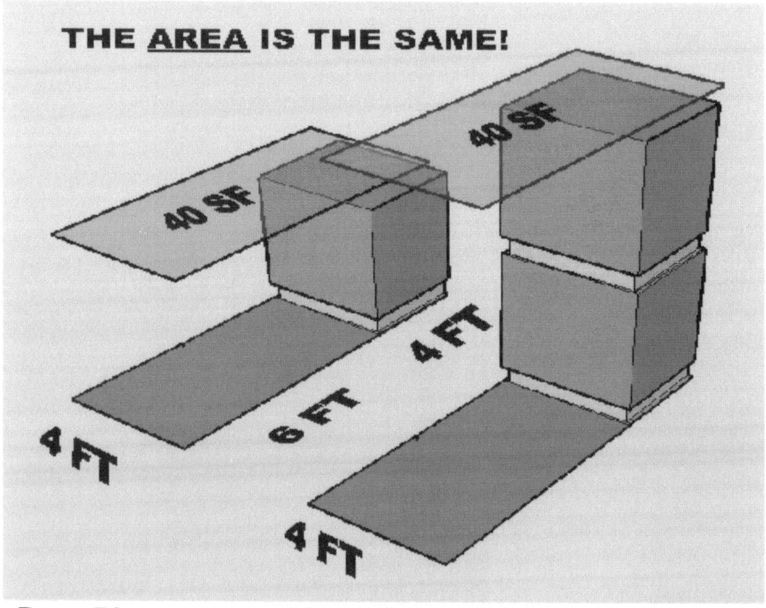

SPACE CHART

EAVE HEIGHT FEET		17	22	27	31	36	41	46.6	51
	TIERS	3	4	5	6	7	8	9	10
EQUIPMENT	AISLE	AREA/LD	AREA/LD	AREA/LD	AREA/LD	AREA/LD	AREA/LD	AREA/LD	AREA/LD
G	5.00	8.99	6.74	5.39	4.50	3.85	3.37	3.00	2.70
AG	5.50	9.32	6.99	5.59	4.66	4.00	3.50	3.11	2.80
F	6.00	9.66	7.24	5.79	4.83	4.14	3.62	3.22	2.90
E	6.50	9.99	7.49	5.99	5.00	4.28	3.75	3.33	3.00
EF	7.00	10.32	7.74	6.19	5.16	4.42	3.87	3.44	3.10
BC	8.00	10.99	8.24	6.59	5.50	4.71	4.12	3.66	3.30
BC	8.50	11.32	8-49	6.79	5.66	4.85	4.25	3.77	3.40
C	9.00	11.66	8.74	6.99	5.83	5.00	4.37	3.89	3.50
C	9.50	11.99	8.99	7.19	6.00	5.14	4.50	4.00	3.60
C	10.00	11.99	9.24	7.39	6.16	5.28	4.62	4.11	3.70
D	11.00	11.99	9.74	7.79	6.50	5.57	4.87	4.33	3.90
D	11.50	13.32	9.99	7.99	6.66	5.71	5.00	4.44	4.00
D	12.00	13.66	10.24	8.19	6.83	5.85	5.12	4.55	4.10

A = TURRET TRUCK. B = STRADDLE TRUCK. C = REACH TRUCK. D = COUNTER-BALANCED TRUCK.
E = TRI-LOADER. F = SIDE LOADER. G = STORAGE RETRIEVAL TRUCK

COMPARE THE RED AND GREEN CHOSEN CASES ON THE SPACE CHART

The green case indicates three tiers of pallets on a 12 foot aisle using a counterbalanced fork truck.
The red case indicates six tiers of pallets on an 8 foot aisle using a narrow aisle reach truck.
The 3 high case requires 13.65 sf per load
For a 2000 load warehouse, the area = 27,300 sf
The 6 high case requires 5.50 sf per load
For a 2000 load warehouse, the area = 11,000sf
THE SPACE SAVING = 16,300 sf
WHAT IS THAT WORTH?

THE COST CHART

The COST CHART is an extension of the SPACE CHART. It is derived by taking the area for each configuration, multiplying it by the appropriate building cost per square area then adding the cost for rack for one load. This yields a cost-per-load stored for building and rack but exclusive of the handling equipment needed. Both charts indicate the types of equipment appropriate for each configuration of aisle and height. Therefore the cost chart on the next page is derived from the space charts by multiplying the AREA per load by the building cost per square foot and adding the rack cost per load. This yields another useful chart that indicates the cost per load for a wide variety of aisle widths, rack heights, and handling equipment types.

It is worth restating that the cost differences between various alternatives are realistic and proportional for your time and place even though the calculations were made in another place and time.

.

COST CHART

MAX LIFT IN INCHES		118	174	230	286	342	398	454	510
RACK COST	$27	$29	$30	$30	$31	$34	$36	$37	$46
BLDG COST/SF	$13.0	$13.05	$14.72	$15.49	$17.99	$19.62	$21.73	$23.75	$26.11
EAVE HT. FT.		17	22	27	31	36	41	45.6	51
TIERS		3	4	5	6	7	8	9	10
EQUIPMENT	AISLE	COST/LD	COST/LD	COST/LD	COST/LD	COST/LD	COST/LD	COST/LD	COST/LD
G	5.00	144.32	128.25	118.95	111.87	109.59	109.26	108.17	115.42
AG	5.50	146.87	131.93	122.25	114.83	112.40	111.97	110.81	118.03
F	6.00	153.02	135.61	125.54	117.86	115.20	114.69	113.45	120.64
E	6.50	157.37	139.29	128.84	120.86	118.00	117.41	116.09	123.25
EF	7.00	161.72	142.97	132.14	123.86	120.80	120.12	118.73	125.86
BC	8.00	170.42	150.33	138.74	129.86	126.41	125.55	124.00	131.08
BC	8.50	174.77	154.01	142.03	132.85	129.21	128.27	126.64	133.70
C	9.00	179.12	157.69	145.33	135.85	132.02	130.99	129.28	136.31
C	9.50	183.47	161.37	148.63	138.85	134.82	133.70	131.92	138.92
C	10.00	187.62	165.05	151.93	141.85	137.62	138.42	134.56	141.53
D	11.00	196.52	172.41	158.52	147.85	143.23	141.85	139.84	146.75
D	11.50	200.87	176.09	161.82	150.84	146.03	144.57	142.48	149.36
D	12.00	205.18	178.73	165.09	153.81	148.81	147.28	145.09	151.95

A = TURRET TRUCK. B = STRADDLE TRUCK. C = REACH TRUCK. D = COUNTER-BALANCED TRUCK.
E = TRI-LOADER. F = SIDE LOADER. G = STORAGE RETRIEVAL TRUCK

COMPARE THE RED AND GREEN CHOSEN CASES ON THE COST CHART

COMPARE THE TWO COLORED CASES FOR COST
The green case indicates 3 tiers of pallets on a 12 foot aisle using a counterbalanced fork truck.
The red case indicates 6 tiers of pallets on an 8 ft aisle using a turret truck
The 3 high case costs $205.18 per load
For a 2000 load warehouse, the cost = $410,360.00
The 6 high case costs $129.86 per load
For a 2000 load warehouse, the cost = $259,720.00
THE COST SAVING = $150,640.00
WHAT IS THAT WORTH?

Assume that 2 trucks would be needed in either Case*:

- 2 Counterbalanced @ $15,000 = 2 x 15,000 = $30,000.00
- 2 Reach trucks @ $20,000 = 2 x $20,000 = $40,000
- Net Increase = $10,000

*** again it is the difference - not today's cost**

Question, would you spend $10,000 to save $150,000?

I WOULD AND SO WOULD YOUR MANAGING DIRECTOR!

In addition, consider the advantage offered by a larger proportion of the total cost being in equipment rather than real estate. Real estate is written off much more slowly than equipment and therefore for the same total investment a much higher cash flow will result.

Chapter 4 : FLOOR STORAGE AND DEEP STACKS

Tip No. 35: **EVALUATE DEEP FLOOR STACKS! (They are not always what they seem)**

THOU SHALT USE THE SMALLEST POSSIBLE AREA. This is one space commandment that is particularly fragile. Aside from the obvious problems caused by such things as skimping on aisles and thus reducing productivity, the really significant effect of this cliche is seen in regard to Floor Stacks. This deep storage method, with its relatively small proportion of aisle, sometimes appears to be more useful than it actually is. The factor that makes it variable is the movement in time of the lots stored. The true efficiency of a storage method starts to change when the first load moves out and is a variable until the slot is ready to be reused. If a single lot is stored in a deep floor slot, then it is obvious that nothing can be stored in front of it without making the original lot totally inaccessible.

The simple fact of the matter is that deep floor stacks are at their highest efficiency at two extremes, one being **A** when the entire lot comes in and leaves as a total movement (full lot shipment), and **B** being the case of dead storage. In between those extremes lies **C** being realistic warehouse movement of varied times of emptying and reusing the space. This gives us a number of situations that bear investigation before a storage decision is made.

A says that the space is completely full and then completely empty and therefore ready for reuse.

B says that the space is completely full for a long time and then moved out. While full, the least area is indeed the best space utilization

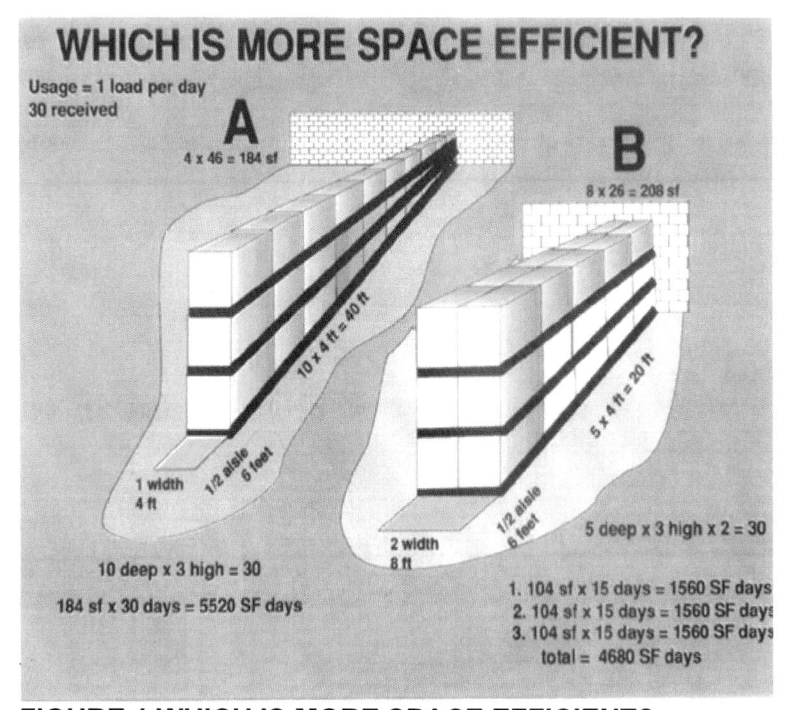

FIGURE 1 WHICH IS MORE SPACE EFFICIENT?

Figure I above shows a common example of the confusion that can be part of storing lots in floor stacks or drive-through rack.

Our example is a lot of thirty loads that are used at a known rate of one load per day. The pallet size is 48 inches by 40 inches, and we have assumed a twelve foot aisle. Since the loads are assumed to be stacked three high,'a floor stack ten loads deep will hold the entire lot, as will two stacks of five deep and three high. A look at the illustration shows that the ten-deep scheme covers 184 square feet as compared to 208 square feet for the two five- deep stacks. The majority of warehousers stop right there and decide that the deeper stack is more efficient because it uses less space. But the calculations shown illustrate the use of "time-occupied" as a true indication of storage cost and efficiency.

We all pay rent on our warehouse space. It may take the form of actual rent paid to a landlord, or it may be an internal cost accounting that shows the cost per square foot for a given time period. Any way you slice it, we pay some dollar figure per square foot per day for the space used in the warehouse. If it is used for storage, it is a good expenditure, and the cost is distributed to a product or activity. If the space is wasted or sits empty, the costs still exist and are not absorbed. If we consider the costs charged to a lot as being relative to the square foot days required, we can evaluate the storage systems in comparable and significant terms.

A further inspection of the calculations at the bottom of Figure I shows that because one side becomes empty, it can be used for other storage.

Case A ties up the entire 184 square feet for the life of the lot, and therefore incurs 5520 square foot days times the rate per square foot per day. Because of the earlier release of part of the space in Case B, the overall cost in square foot days is lower. Is this still true if we use progressively shallower storage? Not necessarily, because it depends also on lot size and movement.

If we look at the example in Figure I and try it with single deep selective pallet rack, we get a total of 6600 square foot days. However, other lotsizes and load movement rates can change this to show an advantage to the rack.

Following are some tables that will help to evaluate various deep stack/movement situations.

FIGURE 2 HOW TO USE THE TABLES		
EXAMPLE: USE THE CASES FROM FIGURE 1		
ASSUMPTIONS: 30 PALLETS IN THE LOT, USAGE OF ONE LOAD PER DAY CONFIGURATION A: 10 DEEP AND THREE LOADS HIGH CONFIGURATION B.: FIVE DEEP AND THREE LOADS HIGH		
	CASE A	**CASE B.**
FROM TABLE 4: AREA PER STACK	184 ft.^{2}	104 ft.^{2}
FROM TABLE 1: NUMBER OF STACKS	1	2
FROM TABLE 2: FACTOR	1	1.5
FROM TABLE 3: TIME TO EMPTY	30 days	15 days

SQUARE-FOOT-DAYS = FACTOR X AREA X NUMBER OF STACKS X TIME TO EMPTY
CONFIGURATION A = (ONE) (184) (ONE) (30) = 5520 FT.^{2} DAYS
CONFIGURATION B. = (1.5) (104) (TWO) (15) = 4680 FT.^{2} DAYS
THEREFORE ANY GIVEN RENTAL PER SQUARE FOOT PER TIME., CONFIGURATION B. IS MORE COST-EFFECTIVE!

TABLE 1 NUMBER OF STACKS PER LOT

AVAILABLE (3 HIGH) CONFIGURATIONS	One deep	Two deep	Three deep	Four deep	Five deep	Six deep	Seven deep	Eight deep	Nine deep	10 deep
NO. IN STACK	3	6	9	12	15	18	21	24	27	30
LOADS IN LOT										
100	34	17	12	9	7	6	5	5	4	4
90	30	15	10	8	6	5	5	4	4	3
80	27	14	9	7	6	5	4	4	3	3
70	24	12	8	6	5	4	4	3	3	3
60	20	10	7	5	4	4	3	3	3	2
50	17	9	6	5	4	3	3	3	2	2
40	14	7	5	4	3	3	2	2	2	2
30	10	5	4	3	2	2	2	2	2	1
20	7	4	3	2	2	2	1	1	1	1
10	4	2	2	1	1	1	1	1	1	1
9	3	2	1	1	1	1	1	1	1	1
8	3	2	1	1	1	1	1	1	1	1
7	3	2	1	1	1	1	1	1	1	1
6	2	1	1	1	1	1	1	1	1	1
5	2	1	1	1	1	1	1	1	1	1
4	2	1	1	1	1	1	1	1	1	1
3	1	1	1	1	1	1	1	1	1	1
2	1	1	1	1	1	1	1	1	1	1
1	1	1	1	1	1	1	1	1	1	1

NOTE: FOR EACH AVAILABLE CONFIGURATION IN THE WAREHOUSE, FIND THE NUMBER OF STACKS NEEDED FOR THE LOT IN QUESTION. USE THIS NUMBER TO FIND A FACTOR IN TABLE 2. ONE DEEP AND DEEP COULD BE RACK. THREE THROUGH TEN ARE MOST LIKELY FLOOR STACKS. ALL ARE ASSUMED AT THREE LOADS HIGH

In between these extremes lie a number of situations that bear investigation before a storage decision is made.

TABLE 2 FACTOR

NO. OF STACKS	FACTOR	NO. OF STACKS	FACTOR	NO. OF STACKS	FACTOR
1	1.0	13	7.0	25	13.0
2	1.5	14	7.5	26	13.5
3	2.0	15	8.0	27	14.0
4	2.5	16	8.5	28	14.5
5	3.0	17	9.0	29	15.0
6	3.5	18	10.0	30	15.5
7	4.0	19	10.5	31	16.0
8	4.5	20	11.0	32	16.5
9	5.0	21	11.5	33	17.0
10	5.5	22	12.0	34	17.5
11	6.0	23	12.5	35	18.0
12	6.5	24	13.0	36	18.5

TABLE 4 AREA PER STACK

CONFIGURATION	AREA PER STACK	CONFIGURATION	AREA PER STACK
ONE DEEP	40 FT.^{2}	SIX DEEP	120 FT.^{2}
TWO DEEP	56 FT.^{2}	SEVEN DEEP	136 FT.^{2}
THREE DEEP	72 FT.^{2}	EIGHT DEEP	152 FT.^{2}
FOUR DEEP	88 FT.^{2}	NINE DEEP	168 FT.^{2}
FIVE DEEP	104 FT.^{2}	10 DEEP	184 FT.^{2}

Assumes a 48" x 40" load with clearances that bring the shadow to 4' x 4', plus one half of the 12 foot aisle

TABLE 3 DAYS TO EMPTY A STACK

NO. IN STACK / LOAD USE PER DAY	3	6	9	12	15	18	21	24	27	30
.05	60.00	120.00	180.00	240.00	300.00	360.00	XX	XX	XX	XX
.10	30.00	60.00	90.00	120.00	150.00	180.00	210.00	240.00	270.00	300.00
.20	15.00	30.00	45.00	60.00	75.00	90.00	105.00	120.00	135.00	150.00
.30	10.00	20.00	30.00	40.00	50.00	60.00	70.00	80.00	90.00	100.00
.40	7.50	15.00	22.50	30.00	37.50	45.00	52.50	60.00	67.50	75.00
.50	6.00	12.00	18.00	24.00	30.00	36.00	42.00	48.00	54.00	60.00
1.00	3.00	6.00	9.00	12.00	15.00	18.00	21.00	24.00	27.00	30.00
2.00	1.50	3.00	4.50	6.00	7.50	9.00	10.50	12.00	13.50	15.00
5.00	.60	1.20	1.80	2.40	3.00	3.60	4.20	4.80	5.40	6.00
10.00	.30	.60	.90	1.20	1.50	1.80	2.10	2.40	2.70	3.00
20.00	.15	.30	.45	.60	.75	.90	1.05	1.20	1.35	1.50
30.00	.10	.20	.30	.40	.50	.60	.70	.80	.90	1.00

The problem with careful investigation is that by the time the calculations have been done, the situation will have changed completely. We have tried to address that problem by developing a simple method to evaluate alternative storage configurations quickly and easily. It is possible for a reader to put this approach on his computer and make the whole process part of the everyday warehouse technique. A far simpler method is illustrated by Figure 2 ABOVE(HOW TO USE THE TABLES).

Knowing the lot size, the usage rate, and the available configurations, we can work our way through Tables 1-4 merely picking out the elements of this simple formula: *SQUARE FOOT DAYS = FACTOR x AREA x NO. OF STACKS x TIME TO EMPTY*

Using this as a guide, it is a simple matter to calculate (or recalculate, as conditions change) the best way to store a lot. Conditions do change, and it's a safe bet that your hard work will quickly become outdated-but it would not be difficult to program this formula for a computer terminal and thus be able to pick the optimum configuration for any lot.

The whole concept of a larger area being more efficient than a smaller one is a minor mind boggler until it is understood. While explaining an example to a client, we were told that the whole thing was theoretical, and that if he moved out the remainder of the lot after fifteen days, the advantage would be cancelled. Our reply was that it was just as theoretical to assume that the lot would be rewarehoused after fifteen days.

This leads to the fracture of yet another commandment: **THOU SHALT NOT DOUBLE HANDLE.** The client was right, and so were we. If the goods were moved at the proper time, it would indeed have been no advantage. The trouble is, very few warehousers consolidate the warehouse on a regular and rational basis. It is avoided as a case of double handling or, at best, done when there is spare time available. Our recommendation is systematically to break this commandment on a planned and regular basis. Tour the aisles on a schedule and instruct the handlers to look for partially full stacks. This can improve space utilization by from 10 to 30%, depending on your product storage mix. In most warehouses that are short of space, the trade-off is such that the handling labor is well spent.

A word to those who wish to adopt this concept for their own use. Any configuration can be reasoned out, and, in fact, you ought to prove to yourself the logic behind the formula. The mathematics of the formula comes from the fact that the successive withdrawals from multiple stacks form a series, and the summation of that series is the "factor." Look for a moment at the bottom of Figure I. Case B shows that for a two stack system, the accounting takes in the whole area plus one half area, or a factor of 1 ½ . A four stack solution would consist of a whole area plus 3/4 area plus ½ area plus 1/4 area = 2 ½ . The examples used to develop the formula also assume a specific load size and a certain number of loads high. These too can be worked out for the particular conditions that you face in your warehouse.

A final comment on the **SQUARE FOOT DAYS METHOD OF ANALYSIS.** The formula can be used to evaluate single and double deep pallet racks against floor stacks and can add a realistic dollar aspect to the comparisons. Some unexpected results will be observed. At first glance, the seductive appeal of deep floor stacks lies in the very low area per load, achieved by distributing the aisle requirements over many loads. This is really theoretical! Walk out in the warehouse and observe first hand how rarely those deep stacks are really full.

Tip No. 36: USE DEEP STACKS FOR LONG TERM STORAGE

If material "lives" in storage a long time, deep stacks are probably best. If the quantity equal to a stack capacity moves in or out all at once, a deep stack equal to the movementb quantity is probably best. For example a stack capacity equal to a truckload when material is shipped by the truickload.

Tip No. 37: USE SHALLOWER STACKS WHEN THE MATERIAL MOVES IN AND OUT RANDOMLY

Analyze the movement with the charts presented earlier. It will probably favor a shallower stack depth.

For those of you who are still with us and would like a more rational and repeatable explanation of why it makes sense to analyze deep stacks, let us present the following:

DEEP STACK ANALYSIS

Since large lots are often stored by palletless methods, and this offers cube and stack height advantages, our discussion assumes the use of a clamp truck or a slip sheet attachment on the fork truck. if stacking height constraints do not inhibit use of the available height or if the pallet
load is extremely stable, then the conclusions reached may be applied to pallet loads as well as to tall columns built by stacking clamped or slip-sheeted loads.

In order to simplify the analysis and approach a general-case solution, we will consider **columns** of storage, **rows** of storage and **stacks** of storage.

These terms (see Figure 1, Definitions) are defined as:

- A **column** is two or more warehousing units tiered one on top of another. This could refer to a number of clamp loads, slipsheeted loads or pallet loads.
- A **row** is two or more columns side by side.
- A **stack** is two or more columns of warehousing units, one behind another.

OPTIMAL STACK DEPTH

An assumption can be made that there is an optimal stack depth which should be used for storing large quantities of unit loads in columns, stacks and rows. The factors on which this depends are:

1. The lot size, i.e., the number of columns of loads in each lot
2. How the goods are released from storage (as a lot, or by portions of the lot over a period of time)
3. Physical turnover of the goods in storage

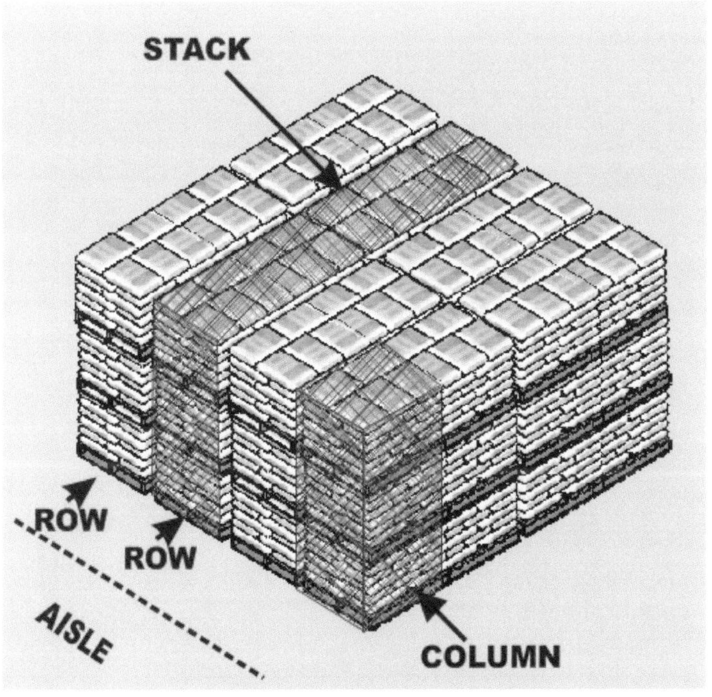

FIGURE 1 DEFINITIONS

Two facts should be kept in mind in order to understand the analysis and to be able to modify it for special uses. The first is that the deeper the storage served by one aisle, the less static space taken. Static space means the absolute floor area at the moment of storage (full slots), regardless of movement of goods or the passage of time. The second fact is that time and movement dramatically affect the efficient use of space. The effect is so great that a larger storage area may sometimes be more space efficient over the life of the storage. A commandment that is particularly breakable is the one that says:**THOU SHALT USE THE SMALLEST POSSIBLE AREA.** This is the basis of the analysis and the conclusions made here.

ASSUMPTIONS

In order to present a general solution that may be used with a degree of confidence, we are presenting comparisons rather than absolutes. By doing this we are saying that such and such is the optimum configuration, that it is better than the others shown. We are not saying that it occupies "X" square feet; indeed, the numbers shown on the look-up table are variously square feet or square foot days, but they too are relative and not absolute. For this reason we can use a real example to establish a useful table for all instances.

SIZES

- 48" x 40" unit load (any reasonable height)
- Three unit loads tiered in a column
- The column footprint is 4' x 4', with clearances
- Aisle width is 12 feet; 1/2 the aisle is part of the stack layout (a stack on each side of the aisle)

MOVEMENT

For Table I, Optimal Slot: First Day of Storage/Long Term Storage

Full lot in on the first day, full lot out on last day of storage.

For Table II, Optimal Slot: For Regular Withdrawal

Full lot in first day, withdrawn at rate of one column per day

For Table III, Average Occupancy Per Day: Regular Withdrawal

Full lot in first day, withdrawn at rate of one column per day

For all Tables

If slot is occupied by any column, then the whole slot must be assumed to be used. This is because a slot would not normally be re-utilized until it is empty. If it were, the original lot would be lost behind the new lot. Furthermore, when a slot is empty it may be used for another lot and is no longer "charged" to the space calculation for the lot in question.

WHY THE TABLES BELOW WERE DEVELOPED

To facilitate the analysis of, and to illustrate what happens in, a multiple stack warehouse, we have developed a series of tables that will yield the correct answers for the lot sizes and stack depths within its boundaries. We have explained the theory in detail so that the reader with different lot sizes may construct his own custom tables.

Table I represents the situation which exists when a large lot is put away and then withdrawn in total. The surface area of each configuration is shown, and the smallest is chosen.

Table II represents a situation where lots are retrieved gradually over a period of time. Based on the assumptions above, the table lists the "square foot days" required for the configuration to empty. Again, the minimum number represents the best choice.

Table III represents the average occupancy of space over the period of storage. The smallest square foot figure is chosen.

Table IV represents the loss in average square feet per day incurred by not using the best configuration. This figure is shown for all of the different ways of storing the lot. This chart may be used as an argument for changing an existing condition.

SAMPLE CALCULATIONS
Consider a lot of 45 unit loads stacked three high to form fifteen columns, all stored at once and removed in total one month later.

Tip No. 38: CALCULATE THE OPTIMUM STACK DEPTH

QUESTION: How to floor-stack for optimum use of space?

Table I FIRST DAY OF STORAGE, indicates on **line 15** that the most efficient storage is in stacks that are 8 deep. The bold number seen is **304**. The dimensions of this module are two load widths (2 x 4 = 8) by 1/2 aisle + 8 depths (6 + 32), or 8 x 38 = **304 square feet.** This is the smallest square footage for storage without
gradual withdrawal. The numbers on this table are square feet.

Table II, OPTIMAL SLOT: FOR REGULAR WITHDRAWAL indicates on **line 15** that the most efficient storage is in stacks that are 5 deep. The bold number seen is **3120.** This represents the number of square feet occupied times the number of days this area is occupied. The storage module is 3 rows of 5 deep stacks. The dimensions are three load widths (3 x 4 = 12) by ½ aisle plus 5 depths

(6 + 20 = 26). 12 x 26 = 312. This is the area charged for the first 5 days, during which the first stack is emptied. The area charged for the second 5 days is 208, the area charged for the final 5 days is 104. If each area is multiplied by the number of days it is occupied and the results added, the sum represents the total square foot days of occupancy. This corresponds to the numbers in the tables.

Table III, AVERAGE OCCUPANCY PER DAY FOR REGULAR WITHDRAWAL is the result of a division of the numbers in Table II by the number of days of occupancy. On line 15, the bold number is **208.** This is the quotient of 3120 (above) divided by the 15 day occupancy. It will be useful to demonstrate the average square feet per day lost by using the wrong configuration. See **Table IV.**

Table IV, LOSS OF STORAGE AREA PER DAY shows the results of subtracting the optimal position value from each of the less efficient configuration values. An interesting exercise would be to multiply the loss per day by the cost per square foot.

TABLE I

OPTIMAL SLOT
FIRST DAY OF STORAGE/LONG TERM STORAGE

NO. OF COLUMNS	1 DEEP STACKS	2 DEEP STACKS	3 DEEP STACKS	4 DEEP STACKS	5 DEEP STACKS	6 DEEP STACKS	7 DEEP STACKS	8 DEEP STACKS
A	B	C	D	E	F	G	H	I
1	**40**	56	72	88	104	120	136	152
2	80	**56**	72	88	104	120	136	152
3	120	112	**72**	88	104	120	136	152
4	160	112	144	**88**	104	120	136	152
5	200	168	144	176	**104**	120	136	152
6	240	168	144	176	208	**120**	136	152
7	280	224	216	176	208	240	**136**	152
8	320	224	216	176	208	240	272	**152**
9	360	280	216	264	**208**	240	272	304
10	400	280	288	264	**208**	240	272	304
11	440	336	288	264	312	**240**	272	304
12	480	336	288	264	312	**240**	272	304
13	520	392	360	352	312	360	**272**	304
14	560	392	360	352	312	360	**272**	304
15	600	448	360	352	312	360	408	**304**
16	640	448	432	352	416	360	408	**304**
17	680	504	432	440	416	**360**	408	456
18	720	504	432	440	416	**360**	408	456
19	760	560	504	440	416	480	**408**	456
20	800	560	504	440	416	480	**408**	456

TABLE I: USE THIS TABLE TO FIND THE OPTIMAL STACK DEPTH FOR LOTS THAT ARE STORED AND WITHDRAWN INTACT AS A GROUP

THE BOLD RED NUMBERS ARE IN THE OPTIMUM STACK. DETERMINE COLUMNS OF TIERED LOADS, READ HORIZONTALLY TO THE BOLD NUMBER, FOLLOW UPWARD TO THE INDICATED STACK DEPTH

TABLE I

TABLE II

OPTIMAL SLOT
FOR REGULAR WITHDRAWAL

NO. OF COLUMNS	1 DEEP STACKS	2 DEEP STACKS	3 DEEP STACKS	4 DEEP STACKS	5 DEEP STACKS	6 DEEP STACKS	7 DEEP STACKS	8 DEEP STACKS
A	B	C	D	E	F	G	H	I
1	**40**	56	72	88	104	120	136	152
2	120	**112**	144	176	208	240	272	304
3	240	224	**216**	264	312	360	408	456
4	400	**336**	360	352	416	480	544	608
5	600	616	**504**	528	520	600	680	760
6	840	672	**648**	704	728	720	816	912
7	1120	896	**864**	880	936	960	952	1064
8	1440	1120	1080	**1056**	1144	1200	1224	1216
9	1800	1400	**1296**	1320	1352	1440	1496	1520
10	2200	1680	1584	1584	**1560**	1680	1786	1824
11	2640	2016	1872	**1848**	1872	1920	2040	2128
12	3120	2352	2160	**2112**	2184	2160	2312	2432
13	3640	2744	2520	**2464**	2496	2520	2584	2736
14	4200	3136	2880	2816	**2808**	2880	2856	3044
15	4800	3584	3240	3168	**3120**	3240	3264	3344
16	5440	4032	3672	**3520**	3536	3600	3672	3648
17	6120	4536	4104	3960	**3952**	3960	4080	4104
18	6840	5040	4536	4400	4368	**4320**	4488	4560
19	7600	5600	5040	4840	**4784**	4800	4896	5016
20	8400	6160	5544	5280	**5200**	5280	5304	5472

TABLE I: USE THIS TABLE TO FIND THE OPTIMAL STACK DEPTH FOR LOTS THAT ARE STORED AS A FULL LOT AND WITHDRAWN OVER A PERIOD OF TIME.

THE BOLD RED NUMBERS ARE IN THE OPTIMUM STACK. DETERMINE COLUMNS OF TIERED LOADS, READ HORIZONTALLY TO THE BOLD NUMBER, FOLLOW UPWARD TO THE INDICATED STACK DEPTH

TABLE II

TABLE III

AVERAGE OCCUPANCY PER DAY
REGULAR WITHDRAWAL

NO. OF COLUMNS	1 DEEP STACKS	2 DEEP STACKS	3 DEEP STACKS	4 DEEP STACKS	5 DEEP STACKS	6 DEEP STACKS	7 DEEP STACKS	8 DEEP STACKS
A	B	C	D	E	F	G	H	I
1	**40**	56	72	88	104	120	136	152
2	60	**56**	72	88	104	120	136	152
3	80	75	**72**	88	104	120	136	152
4	100	**84**	90	88	104	120	136	152
5	120	101	101	106	104	120	136	152
6	140	112	108	117	121	120	136	152
7	160	128	**124**	126	134	137	136	152
8	180	140	135	**132**	143	150	154	152
9	200	156	**144**	147	150	160	166	169
10	220	168	158	158	**156**	168	177	182
11	240	183	170	**168**	170	175	186	194
12	260	196	180	**176**	182	180	193	203
13	280	211	194	**190**	192	194	199	211
14	300	224	206	202	**201**	206	204	217
15	320	239	216	211	**208**	216	218	223
16	340	252	230	**220**	221	225	230	228
17	360	267	242	233	**232**	233	240	242
18	380	280	252	245	243	**240**	249	253
19	400	295	265	255	**252**	253	258	264
20	420	308	277	264	**260**	264	265	274

TABLE III: USE THIS TABLE TO FIND THE AVERAGE SPACE LOSS PER DAY FROM USING THE WRONG CONFIGURATION. SUBTRACT THE BOLD RED NUMBER FROM ANY OTHER IN ITS HORIZONTAL LINE, OR SEE TABLE IV. THESE NUMBERS ARE THE AVERAGE OCCUPANCY PER DAY IN SQUARE FEET.

TABLE III

TABLE IV

LOSS OF STORAGE AREA PER DAY								
NO. OF COLUMNS	1 DEEP STACKS	2 DEEP STACKS	3 DEEP STACKS	4 DEEP STACKS	5 DEEP STACKS	6 DEEP STACKS	7 DEEP STACKS	8 DEEP STACKS
A	B	C	D	E	F	G	H	I
1	0	16	32	48	64	80	96	112
2	4	0	16	32	48	64	80	96
3	8	3	0	16	32	48	64	80
4	16	0	6	8	20	36	52	68
5	19	0	0	5	3	14	32	48
6	32	4	0	9	13	12	28	44
7	36	4	0	2	10	13	12	28
8	48	8	3	0	11	18	22	20
9	56	12	0	3	6	16	22	25
10	64	12	2	2	0	12	21	26
11	72	15	2	0	2	7	18	26
12	84	20	4	0	6	4	17	27
13	90	21	4	0	2	4	9	21
14	99	23	5	1	0	5	3	16
15	112	31	8	3	0	8	10	15
16	120	32	10	0	1	5	10	8
17	128	35	10	1	0	1	8	10
18	140	40	12	5	3	0	9	13
19	148	43	13	3	0	1	6	12
20	160	48	17	4	0	4	5	14

TABLE IV: USE THIS TABLE TO FIND THE AVERAGE SQUARE FEET LOST PER DAY BY USING THE WRONG CONFIGURATION. DETERMINE THE NUMBER OF COLUMNS, READ ACROSS TO ZERO (0) WHICH INDICATES OPTIMUM. THE NUMBERS UNDER THE DEPTH SHOW THE DIFFERENCE IN SQUARE FEET LOST PER DAY.

TABLE IV

CONFIGURATION EXAMPLES

Figure 2, STACK DEPTHS,STACK AREAS illustrates, for a 16 column lot, the appearance of a plan view of each of the configurations. The area per slot or deep row is calculated for each. Note that the static area per column stored decreases as the stacks get deeper.

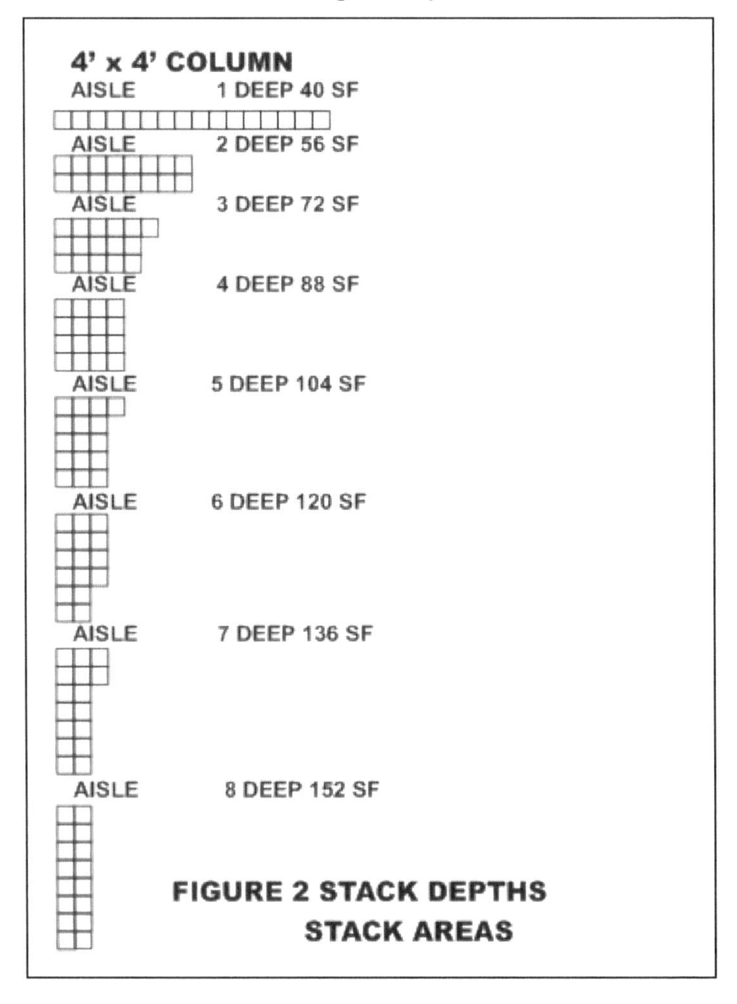

FIGURE 2 STACK DEPTHS STACK AREAS

LOT CONTROL/ HONEYCOMB

When quality control considerations for lot integrity are not required or when strict rotation is unimportant, the loss in capacity caused by partial withdrawals may be compensated for by placing new lots in front of the remaining portion of the old. Such a policy is frequently found in a very high-velocity, high-turnover warehouse, where lots are on hand for only a few days. When strict lot control is required, it is safer to use the reasoning outlined earlier, in the "Assumptions".section.

Honeycomb in a stack is caused by random withdrawals over a period of time. For example, consider a single stack of 5 columns completely full and operating at 100% capacity When one column is removed and another column of the same item cannot be placed in the empty space, the stack will operate at 80% of capacity.As additional columns are withdrawn, the capacity utilization will. decline; 3 Columns left = 60 %, 2 columns left = 40 %, and one column represents 20 %.

From this it can be seen that the average utilization will be 50 % of capacity for a single stack. Because of this, almost any floor-stacking system that uses multiple depth stacks uses more than one row of shallower depth rather than a single very deep stack When lots use multiple stacks, only one stack need be worked at a time. All full stacks over one will be at a height of three or more loads per column, 100% efficiency while waiting their turn to be used up. Thus, any warehouse using multiple stacks is going to be more space efficient than a warehouse in which single stacks predominate.

The tables and theory presented should be used with careful reasoning to make them apply to your operation.

MORE HONEYCOMB THOUGHTS

Tip No. 39: PLAN FOR SOME NECESSARY HONEYCOMB

Unintentional honeycomb is unplanned empty space in a storage or distribution warehouse. Excessive honeycomb is a recurrent example of poorly used space in the warehouse. Some apparently unused space is always necessary for turnaround capability. For example, it is necessary to provide 10% more slots than the average number of loads to be stored simply because at times the day's receipts come in before the day's shipments go out. These figures produce the average, but at certain times more than the average is needed. An obvious place to look for honeycomb is in an area dedicated to the storage of a specific item that is either often out of stock or regularly low in inventory. The honeycomb is that empty space that cannot be used for a different item. The solution to this problem is the rewarehousing of the offending item in a smaller dedicated space. Don't overlook the fact that the rewarehousing itself costs money and affects the tradeoff of money and space. A more effective solution is to install a locator system using floating slots or random storage to warehouse both this type of goods and any seasonal items. Another source of space is found in failure to UTILIZE THE CUBE. Any manager that has not had this maxim drilled into him or her by now is in trouble; in spite of this, we still find many cases of failure to use the available overhead space. Sometimes it can be expensive (a relative term) to capture this kind of wasted space because it may involve the purchase of new equipment, such as trucks and racks, and might require the rearrangement of the entire warehouse.

There are a number of excess honeycomb situations that are not only less immediately apparent but also easier to correct. Sometimes a warehouse manager is actually too

close to his operation to notice these violations. Therefore we would like to point out some things to look for and a way of looking for them that will probably
yield rapid returns.

" MOUNTAINOUS" FLOOR STACKS

Use a warehouse view software simulator or take a ride on the forks of a lift truck (don't tell OSHA!) to a high point in your warehouse. A panoramic view of your floorstacking area will look like a relief map of the Alps. You will see peaks and valleys of stacked height that indicate visually the height utilization of the warehouse. Where potential stack height is four or five loads, you will see one or two loads.

Tip No. 40: STORE SMALLER LOTS IN PALLET RACKS

There will be solid areas of good utilization, completely hiding from a floorbound view as well as rear spots that are either empty or contain low stacks. Often this is the result of attempting to store small lot sizes of the same product in deep floor stacks. At the finish of a lot number, a now stack is started, thus sealing off the rear loads and preventing rewarehousing or consolidation of them.

Tip No. 41: STORE FRAGILE MATERIAL IN RACKS

Stacking constraints are also a contributing factor to the peaks and valleys. Some merchandise can be damaged by the weight of loads stacked above them. Even in a four high area, they may be restricted to a two high stack.
The best solution for lot control and stacking constraints is to store small lots and damage prone loads in selective pallet rack rather than in floor stacks. If only a small expenditure is possible, the damage problem may be solved by the use of pallet stacking frames that allow higher stacking without weight damage.

Tip No. 42: REGULARLY REWAREHOUSE STACKS

Another sight visible from your lofty perch will be very deep floor stack slots with only a few loads at the rear of the slot. Sometimes we fool ourselves when we lay out a warehouse. We figure that 10 loads deep x 4 loads high serviced by one aisle uses a very small area per load stored. This is true only if the slot is full. When it is only partly full, we have, in effect, a thirty foot wide aisle servicing only two or three loads deep; the remainder of the slot cannot be filled with another item, and so we have a great deal of wasted space.

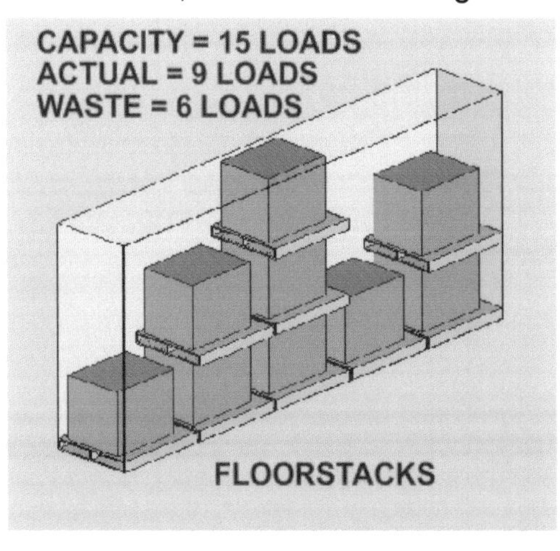

A solution to this is to use different floor stack onfigurations, e.g., shallower ones or regularly and scrupulously to rewarehouse lots that have become too small for their storage positions.

Tip No. 43: **PLACE PARTIAL PALLETS SEPARATELY**

Another deterrent to full height stacking is broken unit loads that have had some cases removed. They are no longer a stable base for stacking, but should not be put at the top of a tier, where they are hard to reach. The next time a few cases are needed, odds are that still another load will be broken into. And if the handler starts another stack in front of the broken load, we still can't get at the loose cases. You probably won't see these loads again until physical inventory time. The solution is either selective pallet rack or a single floor spot across the aisle for case picks. The figure shows a typical floor stack. The capacity of this 5 deep x 3 high stack is fifteen loads, but the actual occupancy is only nine loads. Result: wasted space or Honeycomb.

Tip No. 44: **MATCH THE RACKING AND THE LOAD**

Another source for the honeycomb collector is the improperly assembled unit load that is sized poorly for the rack in which it is stored. The solution is to study the unit loads to see if the height can be increased by adding tiers, or to adjust the load beam of the rack to suit the smaller load. Improper pallet width, or rack that is not matched to the pallet, creates the situation of rack too wide for two loads and not wide enough for three. A change in pallet or rack size is indicated here. The third tier in the illustration demonstrates the solution in which side clearances are

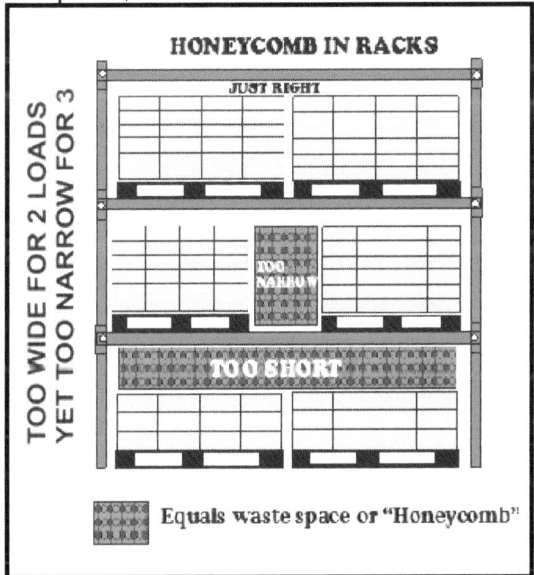

appropriate and the top clearance is only that necessary to lift and move the load.

Tip No. 45: UTILIZE YOUR PALLET BASE

Another subtle waste of space is extreme "underhang" on the pallet. As may be seen in the illustration, there is a lot of

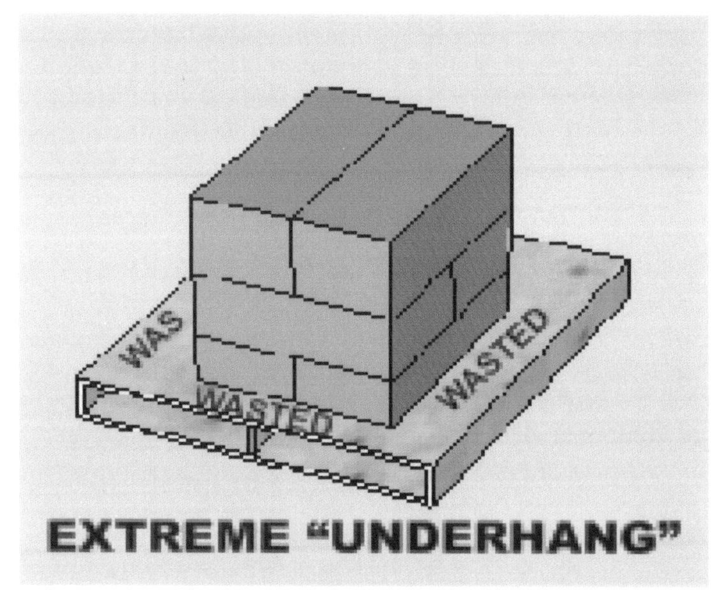

cube loss on the sides of the palletized goodsl. This is not very apparent in practice, and yet it may be the reason that although your warehouse was planned for what seemed to be a sufficient number of loads, it does not hold enough product. The solution is to use the right size pallet and an appropriate pallet pattern.

Tip No. 46: USE THE "RIGHT" STORAGE SYSTEM

Another issue is that of system honeycomb or "built in empty spaces". It is a fact that in order to provide 100 floor stack spaces, you will need about 120 positions. This is because the height cannot always be used for all stock, (stacking constraints), the presence of a single tier where perhaps a lot number change occurred and we want to keep lot numbers segregated, or the fact that stock sometimes goes out more slowly than it comes in and we need a little "turn-around space".

Whatever the reason, there is about a 20% honeycomb factor in floor stacks. Selective pallet rack has about a 10% honeycomb factor because of the constant tier availability, the ability to remove or replace between two tiers and because of the overall accessibility of the stock.

HONEYCOMB FACTORS FOR VARIOUS STORAGE METHODS					
STORAGE METHOD	FLOOR STACKS	DRIVE-IN RACK	SELECTIVE PALLET RACK	DOUBLE DEEP RACK	PUSH BACK RACKS
HONEYCOMB FACTOR	20%	15%	10%	12%	12%

Other storage methods have storage honeycomb factors as shown in the table following. Some are estimates. These estimates can be used to calculate net spaces required for planning a new warehouse or gaining space in an existing one. . For the present remember that to change system honeycomb, change the system.

Tip No. 47: GRAB A BONUS IN HANDLING COSTS

In most warehouses, the cost of handling loads in and out of storage is much greater than the actual storage cost.. A public warehouse, in setting rates for services, pays more attention to the in/out costs than to the monthly storage rate. If it cannot recover both types of cost, it will soon be out of business. Contrast this with captive warehouse distribution centers. The big benefit is in reducing handling costs, and this alone may be the justification for improved load design.

The more units of product that can be handled at one time, the lower the unit cost.. Since the pallet is usually the basic element in handling, pallet efficiency is going to affect cost every time that load is handled. Inefficient use of the bottom load area will increase the number of pallet loads that must be handled in order to move a given amount of product.

Chapter 5: PACKAGING/PALLETIZATION

Tip No. 48: CONSIDER THE BEST PALLET ORIENTATION TO THE AISLE

The size of the pallet and the arrangement of cases on the pallet can greatly affect the use of available space. Consider pallet orientation to the aisle. For example, a 48 by 40, pallet has a width of 48 inches (the second dimension indicates the stringer length) and a depth of 40 inches. This pallet could be stored with the 48 inch dimension parallel to the aisle (A) or the 40 inch dimension parallel to the aisle (B). The space for each load stack can vary dramatically. See illustration below

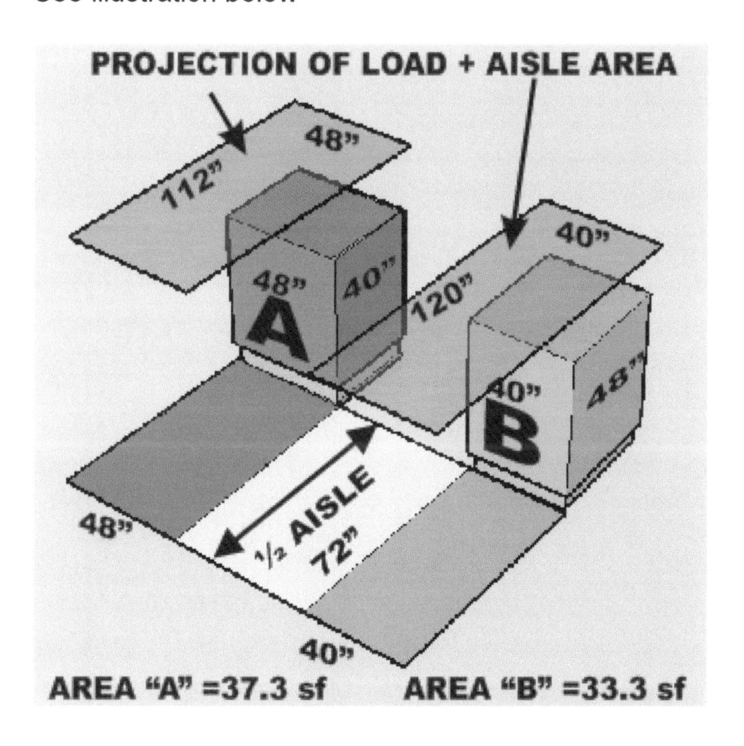

FOR 1000 LOADS, "B" TAKES 4000 LESS SF
The actual choice will depend on the length and width constrictions of the warehouse. Dependent on the type of fork truck used, the aisle width also is influenced by the depth of the pallet. When using rack, the width of the load plus the necessary clearances will affect the footprint of the rack itself. Once again remember that the trade-off between labor and space is a consideration here. As you lengthen the necessary travel you will increase the time needed to select a pallet.

The example shown looks at both a 48" x 40" pallet and a 40" x 48" pallet. This could be the same pallet with a four-way entry. Since the front-to-back and side-to-side clearances are constant, and the aisle constant, the area per load stored will change with the orientation..It can be decided by drawing both out for a final comparison.

When using rack, the width of the load plus the necessary clearances will affect the footprint of the rack itself. Once again remember that the trade-off between labor and space is a consideration here. As you lengthen the necessary travel you will increase the time needed to select a pallet.

Tip No. 49: DESIGN AN EFFICIENT PALLET PATTERN

An inefficient unit load design can not only cause great cube losses, but can also lead to much additional handling. In the example shown below, a 16" x 24" case is placed in a poor pattern on a standard 40" x 48" pallet. This happens because pallet patterns have not been designed for each case used, or perhaps because palletizer personnel have never been told the proper patterns. The net result is harmful: there is excessive hidden honeycomb, handling costs are increased, and you must own more pallets. Consider the same cases when they are in a proper pallet pattern; there is little or no cube or area loss, and you will make 40% fewer trips to move a given amount of product. This can all be achieved through the use of pallet design program (These are computer programs which develop the best patterns for any size case on any size pallet or base.) See Appendix A - “HOW TO DESIGN EFFECTIVE PALLET PATTERNS)

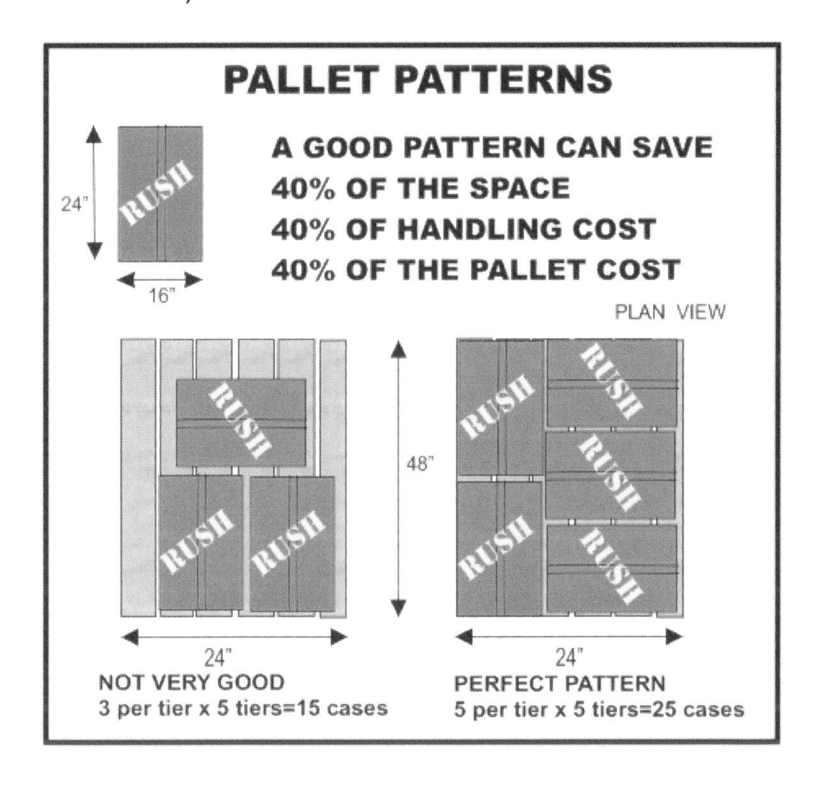

Tip No. 50: **BE CONSISTENT IN PALLET PATTERNS**

As may be seen on the following page, (A PUZZLE: FIND THE MISSING CONTAINERS) inconsistent patterns may increase the difficulty of inventory taking and may affect its accuracy. Shown at the top left are two palletloads that appear identical when viewed from the front, as seen in a warehouse storage slot. Although the two loads appear to have the same tier arrangement, Pallet A actually contains a total of 90 cases, while Pallet B contains only 80 cases. At the right are the pallet patterns for both A and B, with the appropriate code notation for each. (see Appendix "Pallet Patterns, Logic and Theory") Pallet A has the code 3135, which translates to 3 lengths x one row PLUS 3 rows x 5 widths, or *(3 x 1) + (3 x 5) = 18 containers in the tier.* Pallet B has the code 3225, that is, *3 lengths x 2 rows PLUS 2 rows x 5 widths,* or *(3 x 2) + (2 x 5) = 16 containers per tier.*

The 11% loss of utilization from A to B means that an additional 12 pallets per 100 must be used to store the same number of cases accomodated in Palletload A. Perhaps even more important than the additional 12% pallet cost is the problem presented by taking inventory or picking shipments. If Pallet B is assumed to have 90 cases when in fact it has only 80, inventories will be overstated by 11%. Depending on the checking method used, the actual under-shipment may be 11%. In unit load shipments, this type of error is rarely caught.

It is easy to see that with a case of the dimensions in the sample (14" x 10"), it is only too likely that one worker will choose the first pattern while another chooses the second. One of the serious problems in the development of a unit load system is the number of possible arrangements which at first glance *appear* to provide maximum utilization. Diagrams are useful as a way of specifying the load, but the sort of code we describe requires *a physical count* in the very process of interpreting the coded instructions.

Puzzle: Find the Missing Containers

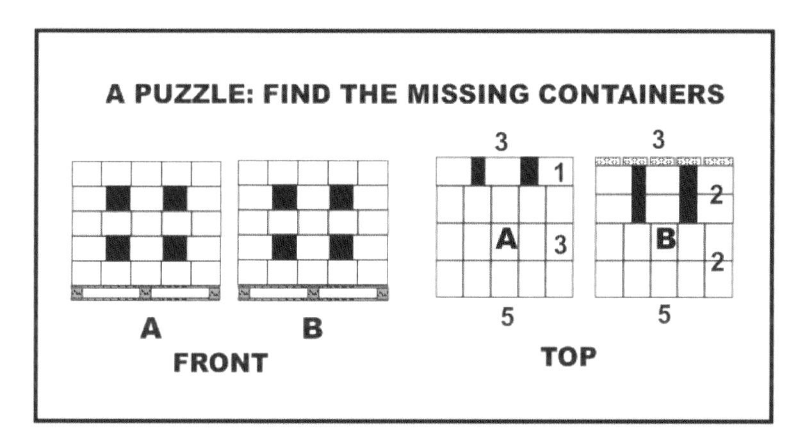

Tip No. 51: ANALYZE YOUR PACKAGING

PACKAGING FOR IMPROVED SPACE UTILIZATION (AND OTHER GOOD THINGS AS WELL)

Several large potential benefits arise from a consideration of the package configuration used in distribution. For example we can achieve better space utilization, reduced handling costs, reduced order selection costs, improved customer service, and even greater sales. see the illustration on the next page and realize that if the cost benefits are claimed for all areas that are improved, the total savings can be huge.

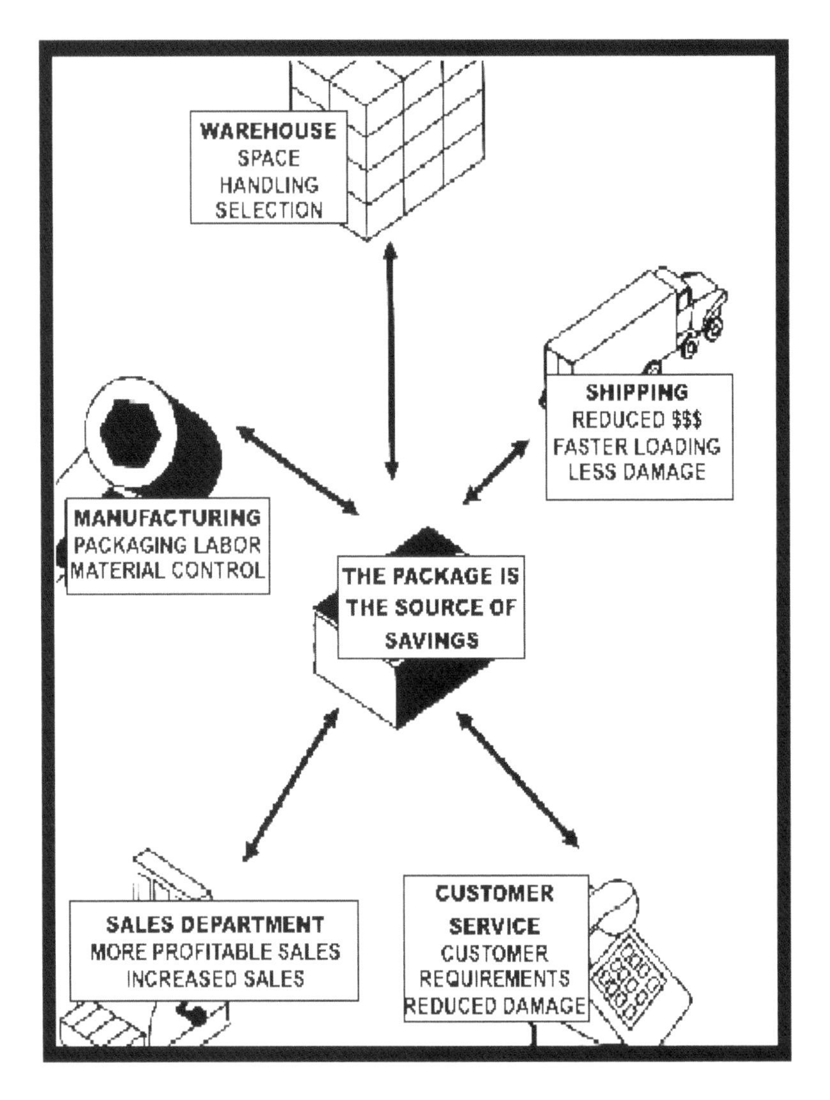

AREAS TO BENEFIT FROM PACKAGE IMPROVEMENT

Below is a checklist to analyze packages for improvement:

✔ CHECK LIST FOR PACKAGE IMPROVEMENTS

SHIPPING

EXPECTED TRUCK SIZE	PACKAGE LENGTH, WIDTH, HEIGHT
TRUCK LENGTH	PALLET LENGTH, WIDTH, HEIGHT
TRUCK WIDTH	PACKAGE WEIGHT
TRUCK HEIGHT	PALLET CUBE (BARE PALLET)
LOAD WEIGHT MINIMUM	PALLET PATTERN

WAREHOUSING

WAREHOUSE CLEAR HEIGHT	PACKAGE LENGTH, WIDTH, HEIGHT
FORK TRUCK LIFT CAPACITY	PALLET LENGTH, WIDTH, HEIGHT
LOAD STABILITY	PACKAGE WEIGHT
PALLET LOAD LIMITS	PALLET CUBE (BARE PALLET)
PALLET STACKABILITY	PALLET PATTERN

SALES DEPARTMENT

CUSTOMER REQUIREMENTS	CUSTOMER ORDER SIZE

MANUFACTURING

PACKING METHOD	BAR CODES?

CUSTOMER SERVICE

BAR CODES?	CUSTOMER WAREHOUSE NEEDS

PACKAGE DESIGN AS IT AFFECTS TRANSPORTATION

Transportation costs may be reduced by:

- Loading more product in a truck at the minimum weight charge.
- Faster loading and unloading of the trailer.
- Less damage because tighter load will minimize shifting.

It is a fact that there is considerable size flexibility in package design. This flexibility is most often consistent with maintaining the packaging goals of product protection. Often there is a "standardization" approach taken to packaging.

This approach closes its eyes to the downstream effects of package design in order to decrease the variety of cartons purchased or other restrictions. An SKU that represents a great deal of activity or inventory should be exempt from such arbitrary approaches and should rather be considered as a unique opportunity to make beneficial and significant changes in the areas of shipping, warehousing, manufacturing, sales and customer service.

By working backwards from each of these areas while respecting the packaging, it is possible to develop dramatic improvements that will favorably affect the bottom line.

AN EXAMPLE

As an example, consider a product that is presently packaged as indicated by the following assumptions.

CASE "A" ASSUMPTIONS

- 8 1/2 inches long
- 7 inches wide
- 9 1/2 inches high
- Weight per case of 5 pounds

The product or products in the case are such that the dimensions could be slightly reduced without harming the protective qualities of the package.

Let us consider an over-the-road truck trailer that has the following internal dimensions

- Length of 40+ feet
- Width of 92 inches
- Height of 105 inches
- Minimum weight charge for 40,000 pounds
- Pallet loading of the trailer
- 48-inch x 40-inch pallets
- Twenty pallets on the floor and two tiers high

The goal in this example is to get more merchandise in the truck. Since the number of pallets is fixed by the floor area, we must maximize the number on the pallet and "utilize the cube" of the truck vertically by storing to greater heights in the available cube. It is of course necessary to leave some clearance at the top to permit handling the pallets with a fork truck onto the second tier.

Tip No. 52: **MAXIMIZE THE PALLET**

A pallet pattern analysis shows us that if we could reduce the 9 1/2 inch dimension by 1/2 inch, an extra case would fit on the long side of the pallet. Similarly, a mere 3/8 inch reduction over it would allow an extra case on the short side of the pallet. (See PACKAGE CHANGE AFFECTS ON SHIPPING) this would fill out the pallet area completely and change the number per layer from 25 cases to 36 cases per tier. Even without adding height, when will that increase the utilization by 44%. This amount will also have a beneficial effect on warehouse handling as well as warehouse storage requirements.

Tip No. 53: **MAXIMIZE THE TRUCK LOAD HEIGHT**

Looking at the truck elevations in that same illustration, we see that the "A" pallet is stacked two high and has a total height including the pallet of 87 inches. The remaining available truck type of 18 inches (105 inches -87 inches) is wasted. We could of course hand stack one additional tier but it is basically impractical to do so. If we can remove 1/2 inch from the individual case it, we'll be able to stack each pallet and additional tier high. The two loads stacked will now total 101 inches. This leaves 4 inches above for easy load movement with a fork truck (105 inches -101 inches). This arrangement adds two tiers of 36 cases on each of the 20 double pallet stacks. Lease for the changes in the pallet pattern and pallet height is an increase of 3200 cases in each truckload.

At 5 pounds per case, the load weight of 36,000 pounds (7200 cases times 5 pounds per case) is still under the minimum load weight and the shipment will travel at the same cost as the case "A" truck load with 3200 fewer cases. Therefore 3200 cases will effectively travel at no cost and the overall transportation saving is 40%.

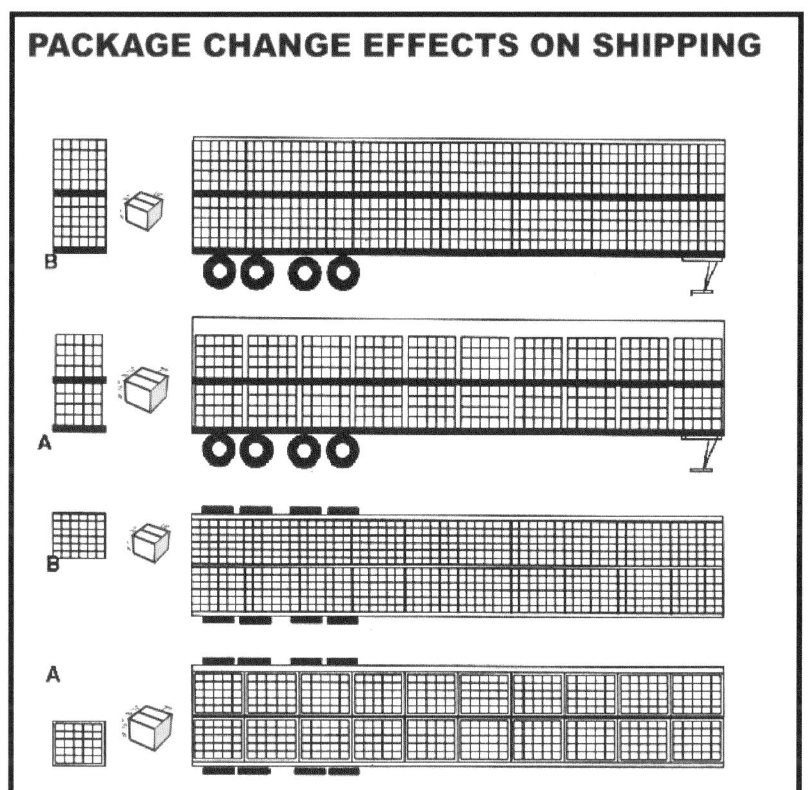

The dimension and pattern for "A." allows 100 cases per pallet and 4000 per truckload. If the product is light in weight, a minimum truck weight will be charged. The dimension and pattern for "B" (smaller dimensions) allows 180 per pallet or 7200 per truckload. Depending on the weight, the extra 3200 cases made travel at no additional freight cost. This represents a saving of 40% in freight for this item. The dimensional changes are slight and often can be accomplished by small changes in package design

Tip No. 54: ANALYZE PACKAGE SIZE AS IT AFFECTS WAREHOUSING

If the warehouse has an 18 foot 4 inch clear height to the sprinklers and we must stay eighteen inches below the sprinkler heads, we have the situation as shown below in "Packaging Effect On Cube Utilization".

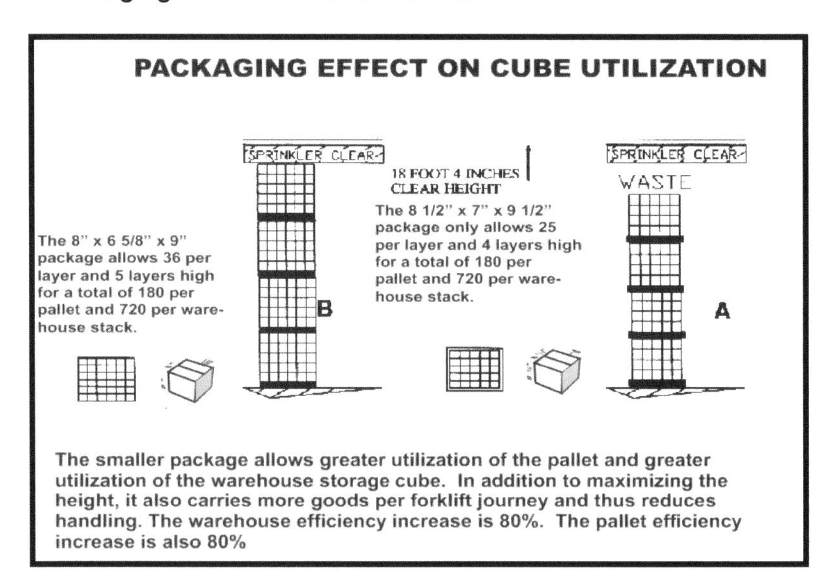

The original "A" pallet is shown at the right side of the illustration. The "A" pallets are stacked four high. We are using floor stacks in the example but similar reasoning would be used for racked merchandise.

"A" stack is 174 inches high including the pallets and thus tops out 45 inches below the sprinklers. If we subtract the 18 inch sprinkler clearance requirement, we are left with 28 inches of wasted or unused height. This is not enough for another pallet load and so is wasted.

The **"B"** type load is also stacked four high but in this case each pallet has five tiers instead of four. The height of this four high stack is 202 inches including the pallets. This height plus the required sprinkler clearance completely utilizes the cube of the 220 inch clear height.

SAVINGS

Each stack plus its half portion of the eight foot aisle takes approximately 32 square feet. In Case **"A"** each case occupies 32/400 = .08 square feet. If the stock turns 4 times per year and the warehouse cost per square foot is $9.00/year, the cost per case for warehouse space is [$9.00/4].08 = $.18. For case **"B"**, each case occupies 321720 = .044 square feet. The cost per case is then [$9.00/4].044 = $.10. The warehouse saving is 44%.

Tip No. 55: PACKAGING AS IT AFFECTS MANUFACTURING

While the package design is under consideration, be aware of the needs of manufacturing. How are they actually putting the merchandise into the packages? Is now the time to consider automatic packing devices or a different closure that will reduce manufacturing labor?

Tip No. 56: PACKAGING AS IT AFFECTS CUSTOMER SERVICE

Input from major customers is part of the analysis; it is possible to accomplish more than one goal at a time. If we can match customer requirements while attending to our improvements, we've increased our value to the customer. Consider their needs and how they use the product and in what quantities. Consider their warehouse needs as well. This sort of information may disclose problems in product protection that can be solved as part of the project.

Tip No. 57: PACKAGING AS IT AFFECTS SALES

The implied savings and advantages to the sales department come through better communication. For example, if the new pallet load capacity represents an increase in the value of a full load, perhaps part of the savings can be passed on to the customer as an incentive to buy in larger multiples or full pallets.

Another saving involved is the advantage of full pallet picks as compared to case picks for an almost full pallet. Creative use of this sort of reasoning can actually increase sales. It has been our experience that the customer is usually most cooperative ... particularly when a saving is involved.

Chapter 6: GENERAL TOPICS

Tip No. 58: ESTABLISH MEASUREMENTS OF CUBE UTILIZATION

Sometimes it is true that merely measuring a situation leads to the solution. Following that line we suggest that you consider the following three indicators of space use.

1.PERCENT CUBE OCCUPANCY
(RATIO OF INVENTORY UNITS X UNIT CUBE) divided by (L x W x H OF THE WAREHOUSE) as a percentage

One very simple measurement (sometimes a little difficult to grasp) is the ratio of total inventory volume (the actual cube of everything stored in the warehouse) compared to the total cube of the warehouse building. By total cube of the warehouse we actually mean the length times the width times the clear height of the total storage area of the warehouse. Realize that this warehouse volume includes aisles, overhead clearance, and honeycomb in the warehouse.

One normally grades any ratio or test on a scale of 1 to 100 where a grade of 90 indicates an excellent result. In this particular ratio, a ratio of 33 percent is considered excellent (again remember that aisles and other clearances are all included. An interesting comparison would be to consider all of your pallets, cases, and individual units stored in the warehouse to be made of ice. Now turn on the heat and let everything melt. In a typical branch warehouse with a clear height of perhaps 18 feet, the water level would probably be less than two feet

2. RATIO OF STACK HT/AVAILABLE HEADROOM
Actual stack height divided by potential height

Here we are comparing the actual height stacked to the absolute maximum we could stack (considering sprinkler clearance of course). This measurement generally indicates how well we are utilizing the height in the warehouse. In this ratio we would hope for a score close to 90 percent.

3. RATIO OF AISLE AREA/OCCUPIED AREA
Aisle area divided by the storage area

This measurement is a two-dimensional measurement in that it compares the area of the aisles to the area occupied by storage. It is useful to see and compare various storage routines. Remember the cube is important as well. In this measurement we would look for a low number.

Tip No. 59: DO AN IN DEPTH ANALYSIS OF YOUR STORAGE

It is possible to improve almost anything by careful analysis. Often, the act of study suggests improvements.
To aid in an in-depth analysis of your storage areas, see on the next page, The "UTILIZATION CHECK SHEET". Tailor it to represent the storage methods and areas pertinent to your situation. Remember to compare the cost of the proposed improvement to the potential savings generated by the improvement. In this way priorities for action will be self-generated.

GOOD AND BAD ECONOMIC TIMES

Although most elected government spokesmen regularly predict that any recession is soon to be over, we can still take advantage of down turns to our advantage. If we take a critical look at the warehouse in terms of a cutback for survival, the increased efficiency thereby obtained will stand us in good stead as the economy turns upward. Tap into the gold mine hidden in your warehouse, and institute an automatic system of improvement inspired by measurement and comparison.

UTILIZATION CHECK SHEET SUPERVISOR ______ WAREHOUSE ______			
STORAGE TYPE	PRODUCT CUBE	AVAIL CUBE	% AVAILABLE UTILIZATION
STORAGE TYPE BULK RACK SHELF BIN FLOW RACK OTHER AREAS RECEIVING SHIPPING OFFICES BATTERY CHG MAINT. LUNCHROOM WASH ROOM			

Tip No. 60: DEVELOP CARTON CUBE DATA

Multiply the length by the width and height for each line item or SKU. Sources for this information include supplier specifications and plain old measurement with a ruler. It would be a good general procedure for receiving personnel to measure any new inbound product. The individual unit cubes must be extended by the inventory quantity on hand.

.

Tip No. 61: MEASURE THE TOTAL WAREHOUSE CUBE

The total cubic space enclosed is obtained by measuring the length, width and height of the area under the roof. This includes offices, ancillary facilities, covered docks and canopied areas. The height includes the total to the bottom of the roof. Do not deduct for sprinkler clearance, etc.

Tip No. 62: **ASSESS THE AVAILABLE CUBE**

A tour of the warehouse is appropriate in order to obtain this data. The analyst should be armed with a scale drawing of the warehouse showing the storage equipment now in place. For bulk storage, it is necessary to estimate the maximum number of levels that can be stacked as well as the number of columns or floor spots. Spots x levels = maximum storage spots. This in turn, multiplied by the cubic feet of a storage spot, yields the available cube in that particular area. Similar calculations for racks as they presently exist yield the rack storage cube. Repeat the exercise for each type of storage area throughout the warehouse. Improvements in storage and layout will be reflected in a higher available cube and a beneficial lowering of the per cent of utilization.

Tip No.63: **CALCULATE THE PERCENTAGE OF UTILIZATION**

Both of the two types of percentage of utilization are calculated by dividing the product cube by the appropriate potential cube. As improvements are made in the available cube without additional inventory stored, the total cube utilization will stay the same and the available cube utilization will drop. As additional goods are placed in the warehouse, both measures will rise. One of the important uses for this measure is in a warehouse that is considered too crowded. If the utilization is small or can be lowered, we know that additional storage is available without outside warehousing or a move to a new warehouse. The availability of such indices is a powerful argument to use for additional space when it really is needed. It is likewise a strong and logical argument to a purchasing department that is prone to surprising the warehouse with huge and unexpected shipments that clog the facility.

Tip No. 64: FOLLOW A ROUTINE FOR ANALYSIS OF CUBE USAGE

Achieving a high cube utilization can increase your profits, make better or more organized use of your own or rental area and, under certain conditions prevent or postpone large capital outlays for expansion of warehouse facilities or rental of outside storage.

The cube utilization analysis program outlined below defines the conditions which may be controlled in a warehouse, shows how to measure and report them and offers suggestions for improvement. Its application will be seen to determine the effectiveness of storage procedures and layout and to increase the accountability of the managers of that warehouse resource. Use of a rigorous technique leads to a rational growth plan and the avoidance of surprises in the realm of warehouse requirements.
Without such a specific and disciplined program, it is expected that the warehouse will have poor cube utilization, crowded facilities, high inventory levels and low turnover, no ability to predict space needs, high labor costs (for stacking retrieval and order filling) and high operating costs in general.

Recognize that there are both controllable and uncontrollable parts of warehouse management. Before tackling the controllable portions of warehouse space management, it would be helpful to present the argument for a modification of outside or uncontrollable influences on the warehouse. It is unfortunate that even today management is often unaware of or indifferent to the fact that many warehouse inefficiencies are the direct result of decisions made by the sales, marketing and inventory management departments. There is no question that the distribution function should service the sales function for that is the basis of a successful company. But those decisions that effect distribution shto ould certainly consider the costs incurred by those decisions. In other words, the bottom line should be the real consideration. Here are some suggestions by influence area that may relate space effectiveness to the outside influences.

Tip No. 65: TIE SPACE NEEDS TO MARKETING AND SALES

In the spirit of the Japanese method of "Kanban" (scheduling parts as needed) , the sales department should provide more frequent and tighter plans and projections of anticipated sales. Of course this involves projecting decreased sales as well as increases, so that stock and thereby space can be minimized insofar as is consistent with good business practice. Although sales do not necessarily dictate the inventory level, it is indeed a function of sales projections.

Tip No. 66: TIE WAREHOUSE COSTS TO PURCHASING

An all too frequent occurrence is the purchase of a special large-quantity "deal" that appears to capture a savings in purchase price. Often, the added warehousing costs (temporary space rental operating inefficiencies due to clogged aisles, etc.) and the carrying cost of the extra inventory, amount to more than the saving in price. If the purchasing department is able to predict or at least to be aware of such costs, they may buy more thoughtfully. Good communication also can have a beneficial effect here.
If the warehouse management is aware in advance of how much merchandise is on order and the expected date of arrival, it can make plans to handle the extra material. Surprises are much harder to accommodate.

.

Tip No. 67: TIE WAREHOUSE COSTS TO INVENTORY MANAGEMENT PRACTICES

It is possible that those responsible for inventory management (quantity levels and order points) should consider storage capacity as an input to inventory level decisions. There are times when the cost of additional storage far outweighs the cost of more frequent purchases and smaller deliveries. It is presently practical to express order quantities in warehouse terms such as pallets or cubic feet and it would be equally easy to compare the available space to that requirement, thus developing the cost of storage that is directly applicable to that particular order or product. Similarly, actions that reduce inventory could be evaluated. An even greater source of savings is the reduction of inventory carrying costs. These annual costs can amount to up to 40 % of the total inventory cost. The inventory management group should also have the authority to dispose swiftly of obsolete or damaged inventory. This will not only save space but will stop the accrual of inventory carrying charges for unsalable goods. Proper application of the above procedure calls for regular reporting of inventory turns by product, expressed in both physical terms (cubic feet) and in monetary terms.

SOME IMMEDIATE ACTIONS

The remedial actions suggested above are of long-term value and will only be implemented with the cooperation of other departments. What can be done internally that will yield immediate benefits? There are a number of physical changes possible that will increase the usable space and many of them are suggested in these pages. It is important to remember, however, that these changes improve the basic storage potential but do not guarantee that the newly gained space will be efficiently utilized.

Tip No. 68: RECOGNIZE THE COST OF SPACE.

It may be a surprise to many that the cost of floor space rarely exceeds 20 percent of the controllable warehouse expenses. In many distribution warehouses, the cost break down is likely to be as follows:

Warehouse labor approximately	60 percent
Clerical and administrative costs	10 percent
Handling and storage equipment	10 percent
Costs of space	20 percent

Although that may seem a relatively small portion of the total cost, recognize that it is second only to warehouse labor. Another influencing factor is that the cost of getting more space by new construction or leasing raises that 20 percent to a very large portion of the total warehouse dollar. It is definitely worth the effort to achieve greater space utilization in order to avoid the additional cost of new construction.

Tip No. 69: SOLUTIONS FOR VERY LOW BUILDINGS

Even in older buildings with a very limited head room it is possible to maximize the storage by considering changing the standard pallet height. The closer to the clear height that can be utilized, the better the space utilization will be. Consider reducing the individual pallet height by the following formula: clear height minus load clearances divided by two. In this way the limited available cube is utilized. Another possibility, but one less acceptable is to hand stack to the limited available height. Here again we may find ourselves on losing side of the trade-off between space utilization and excessive handling.

In the case of shelf storage in the limited height area, the solution could be to install taller shelving to completely fill the available space. It may be necessary to supply a locking ladder to reach the higher shelves.

We were once asked to provide a design for building with a clear height of approximately 11 feet. The activity in this room consisted of the receipt of pallet loads of forms and supplies and the issuing out of packages and singles. The available cube was utilized by using single tier pallet racks with independently adjustable shelf clips on either side. It was not possible to set up a shelf with room above for a pallet of reserve stock.The pallet sizes and load height were chosen to utilize the remaining head room above the shelves . Narrow aisle "walkie" trucks were used to place and retrieve the pallets in aisles also suitable for manual picking.

Tip No. 70: WORK AROUND LARGE COLUMNS AND BAD SPACING

When all else is appears workable in an older warehouse, we invariably find massive concrete columns with a two to three foot diameter. If rack is to be used under these conditions, it becomes necessary to play "the shoehorn game". This of course is nothing new to many of us who use older buildings and new buildings that have been designed without much thought to the contents.

In this case the logical approach is to use a scale drawing of the area with the columns clearly shown in their exact position and size (not often symmetrical) and prepare a clear overlay showing a typical aisle and its accompanying racks. Draw this overlay with varying aisle widths from the minimum for the desired equipment through the ideal up to an oversized aisle width. Choose variable rack lengths and half-sections and mark these on the overlay. The only place that the columns absolutely may not fall is in the aisles. If the columns are in the rack (as opposed to, ideally, falling at the ends), they can be handled in a number of ways:

- Stop the rack short at either a full or half-section and start again on the other side. This will cost you an extra upright or ladder, but it may be worth it.

- Enclose the column with load beams as if it weren't there. The loss may be either one load position or two if luck is against you (Liebeskind's Law: It will be!).

- Compare the alternative layouts and choose the one that offers the greatest utilization of space, provided that the extra rack sections needed do not offset the advantage. As we continually say, "Everything is a trade-off". Similar procedures apply to the layout of shelving.

Shift the aisle overlays back and forth, trying different widths until you get the best placement of the columns with respect to the rack and the best number of aisles. Remember that an aisle against a wall increases the square feet per load greatly; (with storage on one side of the aisle, the entire aisle is dedicated to that one set of loads) so try to avoid that configuration. Also, keep in mind that there is no rule that says all aisle widths must be the same. Sometimes an overly wide aisle or a minimum aisle will make the geometry work out just right. The minimum aisle can be used for slower moving inventory.

Tip No. 71: CHALLENGE AREA WITH LOW COST PER SQUARE FOOT AND HIGH COST PER USABLE CUBIC FOOT

The only solution offered to this shortcoming is to urge the warehouseman to bring the point to the attention of top management in terms of dollars. The cheap rental rate or purchase price of buildings with low clear height, inadequate floor loading and multiple stories is often the only point considered.

The statement often made to the warehouse manager is, ***"You asked for 20,000 square feet; I got it for you. Now let me get my attention back to running the business."*** The counter to this is to calculate the cost per usable cubic foot and to express the warehouse needs in terms of cubic feet required. To this number can be added the additional labor needed to operate in un-satisfactory space. If you don't express your problems in terms of money, you'll never get what you need to operate effectively. Management thinks and makes decisions in terms of money, as in reality they should. Arguments that use subjective terms such as "a better arrangement", "better flow" and "less damage" do not win capital appropriations and top management approval.

Tip No. 72: ANALYZE LABOR COST OF OPERATIONS VS. EQUIPMENT INVESTMENT

The one time expenditure of capital to reduce labor and increase productivity has many advantages. As pointed out earlier, excessive labor costs go on forever, while an equipment investment is made once. In addition, equipment costs can be written off rapidly, while even low building costs are stretched over 30 to 40 years. The rapid depreciation of equipment decreases taxes and causes a favorable cash flow. Also, equipment expenditures are eligible for any investment tax credit that may be available.

All of the above, of course, is in addition to the labor savings or increased ability to handle more business with the same crew. In summation, we can say that it is often possible to improve an old warehouse or a multi-story warehouse by re-layout and by planning the storage techniques used in it.

To those who have one story warehouses with good floors and adequate clear height, we can only say use what you have to its fullest. Utilize the cube by reducing aisles and going as high as you can. If floor stacking height is limited by possible pressure damage to the lower loads, then consider the use of racks. If you are at the point of requiring additional warehouse space, investigate some of the narrow aisle fork trucks that are available. If the conversion to narrow aisle trucks can increase the capacity of your facility by up to 50%, it may be possible to postpone an expansion indefinitely. This space savings will probably more than pay for the new truck! This is an improvement that applies to small operations as well as large.

Tip No. 73: THE PROS AND CONS OF HIGH STACKING

Remember that storage efficiency, including high stacking, outweighs handling efficiency only when space is at a premium. Also handling becomes progressively more expensive the higher you go: more expensive going up; more expensive coming down. When warehouse activity is low, there's seldom justification for high stacking; if you have a three-deep slot and only nine pallets to fill it, why pay the extra cost of stacking a four-high stack and a five-high stack when you can stack three three-high stacks? Here's a slogan to help you remember: "It's not how high you stack it -- it's how many times you stack it that high."

Tip No. 74: CORRECTLY CALCULATE UTILIZATION

Remember that in the typical warehouse there's a big difference -- almost 50% -- between the **capacity** of any given slot or position and the **actual utilization** of that capacity. This is due to the fact that even though the position may be full from time to time, periodic withdrawals are made which leave it only partially filled until additional stock is received. Think of the warehouse that's full at the beginning of the year and empty at the end? On the average, it's half full. So with a typical slot or stack in the warehouse, you can determine **capacity** of the position simply by multiplying the number of rows in the stack by the number of tiers. For example, a four-row by three-tier stack has a capacity of 12 pallet loads. But the **average utilization** of that capacity may be actually six and a half pallet loads (the maximum load of 12 plus the minimum load of one palletload, divided by two to provide an average.

Tip No. 75: WHEN TO SACRIFICE SPACE EFFICIENCY

Without realizing it, you may be permitting some features of your operation to interfere with effective space utilization. But remember that the loss of some space efficiency may be more than compensated for by gains in handling productivity, . and be prepared to sacrifice space when it's advantageous to do so. Here is a checklist of the principal factors interacting with space efficiency:

Ceiling height. Stacking to excessive heights using conventional equipment can be slow and expensive. Many items do not have sufficient package or product strength to take advantage of high stacking. The cost of racks may wash out the space benefits. The exception is an automated storage and retrieval system which can achieve a degree of high density storage rarely practical with other handling methods.

Support column spacing. It may be advisable to lose some space by providing additional aisle width in order to avoid the obstructions caused by columns in the way.

Aisle width. Counterbalanced fork lifts require wide aisles and straddle fork lifts narrow aisles. The added cost of narrow aisle equipment has to be justified by improved space efficiency. Scrimping a foot of aisle width here and six inches there, however, usually creates more liabilities than savings: reduced maneuvering efficiency, and added risks of collisions and product damage or personal injury.

Overall material handling system. In addition to aisle width, other space utilization factors affected by the type of material handling system include utilization of air rights, density of storage, effectiveness of random storage, number of potential facings, and access space.

Pallet size and overhang. Palletization efficiency will be reflected in total storage efficiency. Inefficient pallet patterns, or pallets that do not lend themselves to main product sizes, will naturally result in even less efficient space utilization. As 3M Company discovered, you could end up with as little as 5% utilization!

Unit loads. A standard GMA pallet 48" x 40" x 6" eats up almost seven cu. ft. of space in the warehouse. It does so whether loaded or empty. Slipsheets and also clamp handling offer alternatives that should be explored, even though they may be limited to some extent by the nature of the product.

Forward/reserve system. Forward picking of high frequency items offers significant order assembly economies and may save space with use of a reserve storage area. In general, allocation of space for forward pick and for staging is preferable to excessive travel past the entire pick line merely to select the same items over and over again.

Lot Accountability. In the case of foods, pharmaceuticals and certain other consumer items it may be necessary to completely segregate certain lots from others. In some cases to quarantine product during an incubation period, and in others to assure stock rotation. Unless the lots in question happen to be identical in size to available storage slots, the fact that they cannot be added to existing stocks of the same item in the warehouse inevitably results in further space fragmentation and under utilization. It's one of the hidden costs of maintaining control.

Commodity or **customer grouping.** Regardless of the reason, any proliferation in the number of lots to be stored and kept separate decreases storage efficiency. One reason is of course the additional access (aisles etc.)that is required; this is potential storage space that must be sacrificed, sometimes for relatively low utilization in handling. In addition, overall economies of scale are more and more difficult to achieve as volume decreases in individual lots.

Fixed/floating slot proportion. As discussed previously, an excess of fixed slots is apt to produce serious under-utilization. The general practice is to allocate fixed slots to slow movers and floating slots to high-volume fast movers.

Tip No. 76: MANAGE SPACE EFFICIENTLY

It isn't just the way you lay out the space, it's also the way you manage the operation that contributes to greater or less space utilization. The following questions almost answer themselves -- and they indicate a number of administrative-type steps you can take to improve storage efficiency .

- Do you receive advance notice on all inbound shipments and make use of it to pre-plan space utilization?
- Is a storage layout readily available and in use? Is it used to preplan storage or receipts?
- Do you have a practical location system and are bays and slots clearly marked?
- Are aisle widths being maintained?
- Are pallet racks being used properly? To facilitate order selection? To store small lots more efficiently?
- Is repeated rewarehousing necessary due to excessive honeycombing caused by poor planning?
- Are sufficient pallets on hand? Are they in good repair? Are they of the proper size and type? Are surplus pallets being stored outside the building? Are they protected?
- Is product not needed in the next thirty days stored in closed storage or open storage? Will the use of closed storage provide more aisle facings?
- Is accessibility to merchandise being carried to extremes?

- Is space required for non-storage functions kept at a minimum?
- Is adequate space provided for shipping?
- For receiving?
- Are an excessive number of doors being used?
- Does this consume valuable storage space?
- Do you have a plan to facilitate loading up cross aisles to handle peak storage loads?
- Do you know what your level of occupancy is now?

Tip No. 77: REMEMBER THE COST OF OFFICE SPACE

Managers are often reluctant to include their office space as part of the storage cube. The question is asked,"How does the size of my office have anything to do with the profitability of my warehouse?" When a rental fee for office footage is added to the departmental budget, the manager quickly finds a way to reduce it. For example, when the office is an eight-foot-high cube built in the middle of a twenty-four-foot clear warehouse, the idea of double decking (building a mezzanine above the office) becomes very attractive. In such a situation, one manager turned a rarely-used conference room into a small-parts storage area, another moved the office to a mezzanine and stored small parts below. The moral of the story is that offices should be counted as part of the total cube. This is particularly true when the prime function of the building is warehousing and distribution.

Tip No. 78: LOOK AT YOUR PRODUCT MIX

Determine the relative importance and activity levels for various products. Recognize that some items live in the warehouse a long time and some a very short time. (A low number of inventory turns versus a high number of turns). Those items that are low turns take a relatively large amount of space for given amount of sales. Conversely, fast-moving items take a relatively small space for a given amount of sales. If you're inventory is heavy with low turns items, consider dropping those from your business plan or storing them in configurations that require less space and greater handling. The greater handling is sometimes a good trade-off for saving a great deal of space.

Tip No. 79: GET RID OF OBSOLETE INVENTORY.

For example many organizations have a returned goods area where returned goods are storage until a satisfactory disposition has been chosen. It is our experience that the disposition of such goods often takes excessive time and therefore the material takes up space longer. Is greater than the value of the space to be gained by disposing of it. How do we dispose of this excess inventory?

1. Refurnish the inventory by fixing any of the problems that caused it to be returned.
2. Repack or repackage in order to match current inventory.
3. Return to the manufacturer for credit or replacement
4. Some of the goods to a in "closeout dealer" to salvage part of the cost and reduce the space taken.
5. Export the goods to a 3rd World country
6. Take the material apart in order to salvage good parts or recyclable material.

7. Scrap the excess material in order to release the space for useful purposes.
8. Make a donation to eligible charities and take a tax write off.

Tip No. 80: INSTALL OR IMPROVE YOUR PRESENT STOCK LOCATOR SYSTEM

The fundamental goal of a stock locator system is to make the best use of your valuable storage space. If you're system results in a large percentage of "honeycomb" (partially filled storage slots and blocks) then you should re-examine it and look ie alternative storage methods including fixed vs. floating slots. A certain amount of honeycombing is inevitable because of the intermittent nature of warehousing (an empty spot must exist in order to accept incoming loads) but beyond a certain point it gets so expensive that you must rewarehouse, i.e., consolidate stock so that will take up less space. Note that this represents an unproductive double handling. Once again the deadly trade-off enters the reasoning.

Tip No. 81: CONSIDER FIXED SLOTS VS. FLOATING SLOTS

A warehouse in which all storage locations are fixed or dedicated is generally quite inefficient because of the amount of space that must be dedicated to inventories that fluctuate considerably in size. For example we must have dedicated space for material that is not yet arrived. A storage slot dedicated to 8 pallet loads might have an average occupancy of something like four or five pallet loads because goods are withdrawn to a certain level before the slot is replenished. Multiply this type of inefficiency by the number of items stored and you can see why some compromise is needed to achieve a reasonable use of space.

Tip No. 82: COMPROMISE FIXED VS. FLOATING SLOTS

Here is one way that such a compromise works. High volume movers can be stored in floating or random slots and therefore order selectors are required to look up where those items are located (look up in a locator system). Medium movers can be stored in fixed locations with reserve stock stored at random in the same general area. Dead inventory or extremely slow movers can even be stored in back of faster movers with a letter R to denote a rear position.

Tip No. 83: INCREASE TURNOVER

Consider these points before you go all out to increase turnover. The optimum is achieved when maximum dollar sales are achieved with minimum investment in inventory.

Increase turnover by:

- Reducing inventory level,
- Increasing sales and promotion using the same stock,
- Increasing variety

Gain these advantages:

- Use capital more efficiently,
- Free up capital for other uses
- Minimize risk of obsolescence.

Incur these risks:

- Surrender of quantity discounts,
- Increased freight costs,
- Increased warehousing expense,
- Lost sales due to excessive stockouts,
- Lost sales due lack of variety.

Chapter 7: RECEIVING AREA

The receiving area is often ignored from the standpoint of increased space utilization. Nevertheless it can yield greatl opportunities to create space that you never knew you had.

Tip No. 84: **UTILIZE THE CUBE**

You will see this one repeated in every warehouse functional area. Simply stated, it is the key to finding more storage space.

Tip No. 85: **ADD RACKS**

The introduction of racks where none exist can drastically improve available space where insufficient exists for items that normally accumulate on the floor. In receiving, their use can facilitate an improved incoming staging capability..

Tip No. 86: **COMBINE SHIPPING AND RECEIVING**

In most facilities, shipping and receiving peak at different times. If the dock locations allow it, one staging area could serve both functions. One combined office would also serve both activities.

Tip No. 87: **LIMIT THE RECEIVING STAGING AREA**

Save space in a receiving area by limiting the size and area to a calculated one to two days of receipts. Not only will material reach the stage where it can be selected and shipped to the customer faster but the incoming staging area will be under control and therefore smaller.

Tip No. 88: **SCHEDULE INCOMING TRUCKS**

By spreading the arrival of trucks, it is possible to reduce overload and jams at the docks and the basic unload area. Please believe that you **CAN** schedule incoming trucks.

Tip No. 89: FILL INTERIOR TRUCK DOCKS

Another source of additional space can be found in those facilities that have inside truck docks. Filling in those docks and then using an outside truck door with a weather seal and canopy can often provide enough space to avoid a move to a larger warehouse. Most inside docks are about 15 feet wide and 65 feet long. The area to be picked up is 975 square feet, enough for up to 148 additional loads, depending on the storage system used

Tip No. 90: ASK THE PURCHASING AGENT TO ESTABLISH RECEIVING SPECS

If the purchasing specifications designate a particular pallet pattern for received goods, we will no longer need extra space staging layout and re-palletize inefficiently palletized goods. Please note that not only is this a space saving but will also save a great deal of labor.

Tip No. 91: PLACE RACKS ABOVE THE DOCK DOORS

Racks above the dock doors can hold empty pallets or they are most needed and thereby free up prime storage areas for merchandise. This also places empty pallets or they may be most needed for the receipt of on palletized material.

Tip No. 92: LAND USE

Zoning restrictions, setbacks, and rights of way can often be utilized as temporary storage or staging and even as temporary truck parking while waiting for a dock.

Tip No. 93: STAGE INCOMING PALLET LOADS ON RACKS

If there is a shortage of staging area, consider racking some of the palletized staged material. By utilizing the height in that area more material can be staged. Recognize that you will incur a slight increase in handling.

Tip No. 94: STORE EMPTY PALLETS IN INEFFICIENT RACK POSITIONS

A "poor" rack position occurs when the clear height of the building is such that the top rack shelf cannot accommodate the height of a full pallet load. A short stack of empty pallets is a good way to utilize that space.

Tip No. 95: ASSIGN ADDRESSES TO STAGING SPOTS.

This will allow you to make receipts more swiftly available as shippable inventory. In many cases it will empty the staging area more rapidly. This works best if you have a real-time computer system, because the staged inventory is entered into inventory rather than going into limbo until it reaches its final storage destination. Grocery warehouses do this all the time, since they regularly schedule shipments of material for the day it is to be received. It is particularly important in that business, since there are no back orders, and an unshipped sale is a lost sale.

Tip No. 96: OUTLINE AISLES AND STAGING

Spend a hundred dollars for a line painter, and outline storage areas and aisles. (Special tape may be purchased instead.) It has been proven that the tendency to leave goods in the aisles or staging areas is reduced when they are clearly delineated

Tip No. 97: PLASTIC PACKING PEANUTS

If you use a lot of this stuff, consider building a storage hopper above the warehouse in order to be able buy it in bulk. Load the hopper by vacuum hoses from a freight car. There are amazing savings when this material is bought in bulk. In addition, large bags will not take up floorspace.

Tip No. 98: ANALYZE YOUR STAGING AREA

Ask yourself the following questions:
Is the staging area completely filled? If so, it may be too small. Is the staging area relatively empty? It may be too large and lead to excessive cycle times waiting to be put away. If a staging area is provided for three days receipts, material **will stay there as long as three days**. When only one day's staging is provided, efforts will be made to move goods into storage without delay in order to free the staging area. This is a good thing!

Tip No.99: TRY FOR JUST IN TIME (JIT) WITH YOUR SUPPLIERS

Even though you are not Toyota and cannot dictate unilaterally to your suppliers, one can still work with the suppliers to help approach Just-In-Time service.Smaller receipts require less space and there is pressure to get them in the system as soon as possible.

Tip No. 100: USE CROSS DOCKING

If incoming material is needed for distribution to a number of predetermined destinations, it makes good sense to ship it out immediately without putting it away. Taken directly from receiving to the outbound docks. This will save space in the receiving area and reduce handling as well

.

Chapter 8: STORAGE AREA

Tip No. 101: REDUCE AISLE WIDTH

The purchase or lease of narrow-aisle handling equipment and a subsequent re-layout will result in more usable storage for a given area.

Tip No. 102: USE MULTI-DEEP SLOTS

Consider the use of two- to four-deep slots (drive-in, drive-through or doubledeep racks or floor stacks, if applicable to the lot sizes and stacking constraints) for high turnover and bulk product.

Tip No. 103: USE TALLER LIFT TRUCKS

Acquire (buy or lease) fork trucks that can store to greater heights. The cost will be minor compared to buying, building or leasing additional space.

Tip No. 104: STORE IN TALLER RACKS

Acquire new and taller racks or add height extensions to existing rack in order to increase the usable cube. Once again, the cost will be minor compared to buying, building or leasing additional space.

Tip No. 105: ARGUE WITH FIRE UNDERWRITERS

Argue with fire underwriters about height restructurings. Often they will negotiate and sometimes "back down" when faced with chapter and verse of precedent or what others in your area are doing.

Tip No. 106: FIGHT WITH SAFETY ENGINEERS

Their recommendations are often arbitrary and conservative. See what others are doing in your area. Compare and argue for higher storage.

Tip No. 107: **SERVICE AREA MEZZANINES ABOVE STORAGE**

Consider putting offices, lunchrooms, and washrooms on a mezzanine with low-turnover stock or small parts beneath.

Tip No. 108: **BRIDGE THE ENDS OF AISLES**

By placing load beams and cross members across the end of the aisle, it ios possible to gain several additional pallet positions. In the illustration below,

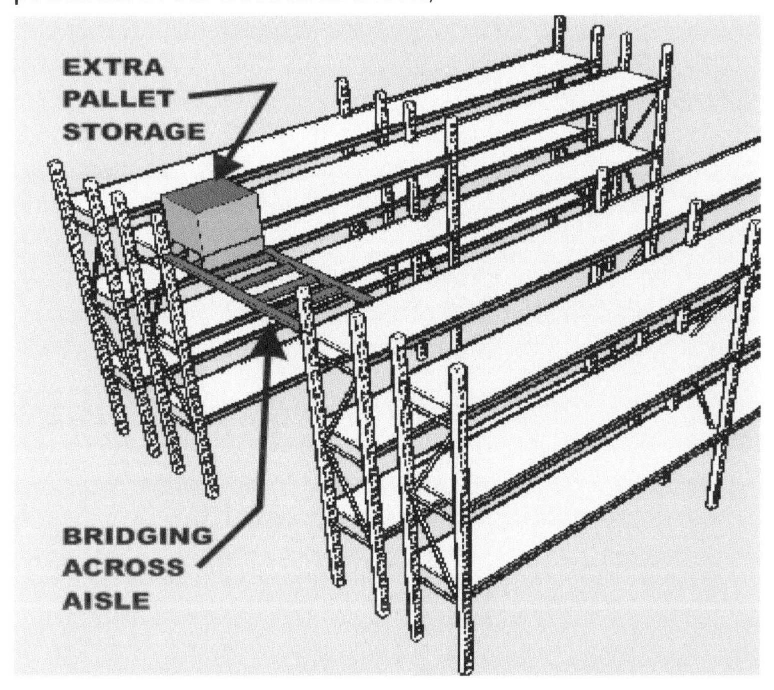

Three pallets could be stored on the bridge beams. Access would be lost for one pallet on either side of the aisle at the end. That means a loss of two and a gain of three for a net increase of one additional pallet load.

Tip No. 109: USE JIT FOR CORRUGATED CARTON SUPPLIES

Take advantage of the fact that there are a great many suppliers of corrugated cartons. So many in fact that to be competitive, they will regularly deliver for no charge, small quantities sufficient for only several days at a time in exchange for a decent order. They will store the cartons free and you can use the space formerly taken by the storage of large supplies of packaging.

Tip No. 110: USE DOUBLE DEEP RACKS

Where pallet inventories are such that there are many instances of more than one pallet identical to another it can be useful to go to double deep racking where two pallets are stored one in front of the other. This allows one aisle to serve two pallet loads on each side. This can also be used with unique pallets but at the cost of additional handling. This indeed may prove to be a worthwhile trade-off depending on the cost of space and the cost of handling.

Tip No. 111: SWITCH FROM FLOOR STACKS TO RACKS

Utilize the full height of the available storage area by switching from floor stacking to the use of pallet rack. Where stacking on the floor might be limited to 2 loads high (because of the crushing weight of the loads above or by instability of the load base) you will be able to stack to the full clear height without excessive pressure on any pallet.

Tip No. 112: USE PUSHBACK RACKS

Similar reasoning applies to such innovations as push back rack. This rack can have a depth of from 1 to 4 pallets of the identical product. As the front one is removed, loads in back roll forward. When the new load is put in it sits in the front and is pushed back. This is repeated until the rack is full. It allows one aisle to serve multiple depths of storage.

Tip No. 113: USE DRIVE IN RACKS

Drive-in and drive-through racks offer similar use of the aisles. Be careful to consider the actual occupancy of these racks over time (see the discussions relative to floor stacks as it also applies to these deep occupancy racks)

Tip No. 114: ADD A MEZZANINE

Another way to pick up space in a modern small warehouse is to add a mezzanine storage (or pick) area over an existing pick area. The space above is usually wasted, and the cost per square foot gained is very much less than the cost of additional space. A landing platform or drop point can be provided at the edge of the new mezzanine so that fork trucks can drop off replenishment loads and thus avoid the necessity of purchasing a vertical lift.. Again, good planning of storage can place the slower moving goods on the mezzanine level and thus avoid additional travel.

Tip No. 115: CONSOLIDATE DIFFERENT ITEMS

It is *possible* to consolidate different items on the same pallet. But be very careful doing this as it offers a possibility of errors in selecting the items. One way to reduce this problem is to give a different address to each half of pallet and make sure that the addresses are clearly shown.

Tip No. 116: **ESTABLISH A DIALOGUE WITH YOUR PURCHASING AGENTS**

Encourage them to establish truly appropriate levels of inventory. If stock is used up more quickly, the net space is reduced. It is a fact that often purchasing agents will make huge purchases in order to save a small amount in the purchase price. In many cases the overstock takes up space that is more valuable and more costly to replace than the savings made by the purchase discounts.

Tip No. 117: **"BILL" THE SALES DEPARTMENT**

Consider rendering suspense "bills" to the sales department for excess storage of slow-moving inventory. Every warehouse hides a few mistakes, and it is often the final resting place for that inventory. Making sales aware of their "dirty little secret" will often trigger disposal of those goods. *The "bills" are really "suspense accounts" that disappear without affecting the real cost accounting.*

Tip No. 118: **USE A SYSTEM THAT RECOGNIZES THE REALITIES OF HANDLING SMALL LOTS.**

Flow-through storage for small lots saves access space whenever you can put one incoming lot behind another without having to handle and store the new lot through a reserve storage system.

Tip No. 119: **USE RANDOM STORAGE**

The use of a stock locator system can help accomplish the achievement of the maximum utilization of available space. random storage (anything may be stored anywhere) is the best means of obtaining maximum space utilization, but remember that a random storage system must be supported by good stock location control, including marking of all warehouse bulk storage areas and keeping accurate locator records. A change from dedicated storage to random can increase storage capacity by up to 33%. The basic reason is that dedicated storage holds space empty to accommodate material that is not there.

Tip No. 120: **USE SPECIFIC ADDRESSES**

An address must guide the putaway or retrieval operator to exactly the spot where the desired goods “live”. General areas leave too much to the memory of the operator. A good locator and address system is a necessity for random storage.

Tip No. 121: **GET RID OF HONEYCOMB**

As in all other parts of the warehouse as well, watch for rack spots and floor spots that are not occupied by stock but are not usable. Watch for rack position heights and widths and unused clear height in the warehouse. Also be aware of package and pallet designs as they affect space utilization.

Tip No. 122: **MANAGE SPACE PERFORMANCE**

Measure and track the percentage occupancy numbers for your storage areas and as they change, be aware of the reasons for change and act to constantly improve that performance. Any changes will immediately be reflected on a report and can be clearly identified as an increase in available cube or a change in product storage level. It is recommended that the reports and graphs be generated weekly although monthly reviews can be effective in tightly controlled companies. The important thing is that beneficial changes be recognized and perpetuated while fixing negative changes.

Tip No. 123: **GET RID OF JUNK IN THE WAREHOUSE**

Junk can be defined as obsolete office furniture that the office manager could not dare to throw away. It can be a dozen palletloads of 20-year-old accounting records. Junk can be obsolete merchandise that was written down two years ago and can be the boss's personal furniture from his last career move. (Maybe even his personal boat). Do not get carried away in this task, particularly about the boss's storage, but in the average warehouse one can find a great deal of space that is presently dedicated to the storage of junk. It takes fortitude to cart that off to the dump but it beats hiring outside storage for current stock.

Tip No. 124: **REARRANGE TO AVOID OBSTRUCTIONS**

When building columns obstruct an aisle it is sometimes necessary to move the rack so that the offending column is inside a load space. Better to lose a few loads than obstruct the entire aisle . The load loss will be the same in both configurations.

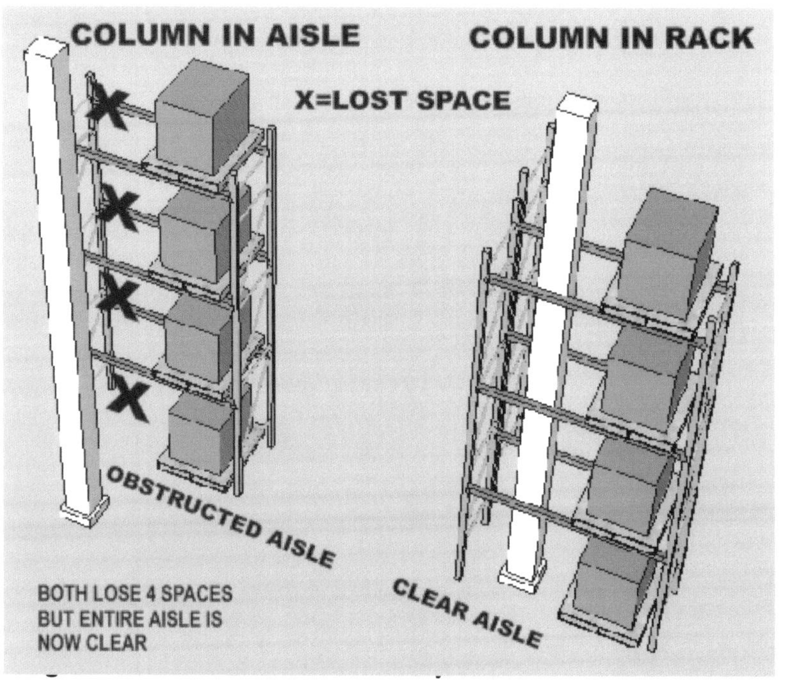

Tip No. 125: MOVE THE LOAD BARS ON PALLET RACKS

Or even purchase additional bars so that you can store more levels of the shorter height loads that appear in your warehouse. The important thing to remember is that a pallet or other unit load takes the same space whether it is full or half empty.

Tip No. 126: CHECK THE LOAD BEAM LENGTH

Compare the length to the pallet width. Three to four inches between pallets is sufficient; more is wasteful. Often the racks were purchased to accommodate a different size than the one now in use, or were purchased at auction with no attention paid to the goods to be stored. Think about justifying new rack or a different pallet size in order to utilize more fully the space in your warehouse.

Tip No. 127: MAKE A SCALE DRAWING OF THE WAREHOUSE

Develop a scale drawing of the warehouse showing aisles, racks and storage, as well as low hanging obstructions to vertical storage. A shocking number of warehouses do not have, and never have had, an accurate layout of their storage. How can you accomplish a plan if you can't see the overall picture? How can you decide on pick paths and cross aisle requirements without an overall idea of the relationships among storage areas?

Tip No. 128: **RE-LAYOUT**

Now that you have your map of the warehouse, use it as an excuse to see how much more storage might be made available through the use of narrow-aisle storage. It is not uncommon to achieve a 30% greater utilization through the use of reduced aisles. Be careful, however, not to overdo aisle reduction to the point where ease of storage and retrieval suffers.

Tip No. 129: **OUTLINE AISLES AND STAGING**

Just as in other warehouse areas where goods are stacked or racked, paint the lines that indicate the boundaries and outlines of storage. The integrity of the aisles will be dramatically improved.

Tip No. 130: **CONSOLIDATE PARTIAL PALLETS**

Note the instances where partial pallets of the same item exist, each in a full load spot. Each partial pallet takes up a full space and every pair that is consolidated releases a full spot for storage. Rare is the warehouse that does not have hundreds of spots that could be released by regular consolidation. If you find these trips tiresome, consider the use of a Warehouse Management System for controlling inventory and storage. A WMS can generate a list of partial pallets automatically with a list of locations and the quantitiesfor consolidation.

Tip No. 131: **USE AISLE GUIDANCE FOR VERY NARROW AISLE TRUCKS**

The turret truck, by virtue of its swinging forks or mast, allows a much narrower aisle to be used. If you have very narrow aisle turret trucks utilizing, for example, a 6-foot aisle, the aisle can be made narrower still if an aisle guidance system were to be used. This can be a signal wire in the floor or the familiar steel channel guides at either side of the aisle. In a very large warehouse, this reduction in aisle width could result in the addition of an entire new section of rack and the resultant additional storage space.

Tip No.132: **USE CANTILEVER RACK FOR LONG AND BULKY MATERIAL**

Place large or long and bulky goods on cantilevered racks with or without plywood decking. This will use air space instead of floor space. The cantilevered racks can be served by an overhead crane or by special in-aisle cranes or a special duty fork truck. The racks relieve the load length restrictions of selective pallet rack.

Tip No. 133: **USE A MOVING AISLE RACK SYSTEM**

These systems mount the racks or the shelves on wheels that ride on a steel track. There is only one aisle and that is movable as the racks or shelves slide to one side or the other either manually, or powered. The savings can be seen in the illustration below.

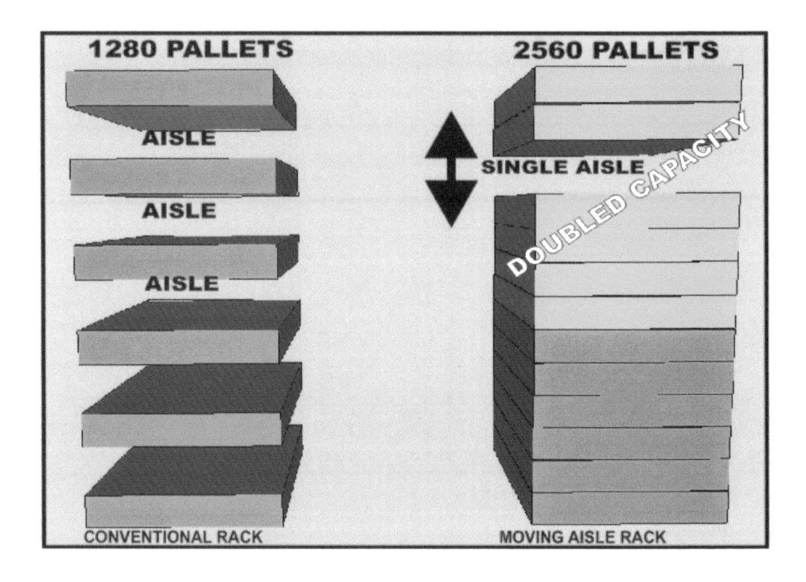

Keep in mind that this is an expensive way to gain space. Sometimes there are situations that justify the expenditure. Similar solutions can be used for small items on shelves.

Tip No. 134: ESTABLISH JIT (JUST IN TIME) WITH YOUR SUPPLIERS

Work with suppliers to help approach a Just-In-Time supplier relationship. The secret is to clearly and fairly indicate the requirements and to set a price that is acceptable to both you and the supplier. Further work to establish quality confidence such that eventually no incoming quality control will be needed. You must also let your supplier be privy to your production so that he will know projected needs. Many partnerships share a computer link so that the purchaser and the supplier have free access to each others MRP or production plans. AS mentioned earlier, this also reduces space pressures on the receiving area of the warehouse.

Tip No. 135: REWAREHOUSE FREQUENTLY

If you're storing in deep stacks, frequently check to see that the various slots are full to the front if appropriate. Also make sure that for a given item, only one stack is used. If more than one stack is active and the items are identical, consolidate them in one stack That will reduce the honeycomb and allow a new item to be stored in the completely empty stack.

Tip No. 136: REGULARLY USE "MANAGEMENT BY WALKING AROUND"

Regularly stroll about the warehouse with notepad in hand to note instances of possible load consolidation, unnecessary honeycomb, partially empty deep stacks and other sources of opportunities to increase space utilization in the warehouse.

Tip No. 137: CONSIDER CROSS DOCKING

If incoming material is needed for distribution to number of predetermined destinations, it makes good sense to ship it out immediately without putting it away. In other words the material goes directly from the inbound to the outbound docks therefore it is called cross docking. The practice avoids the use of regular storage space to hold the material before picking it again. The practice not only saves space but reduces handling as well.

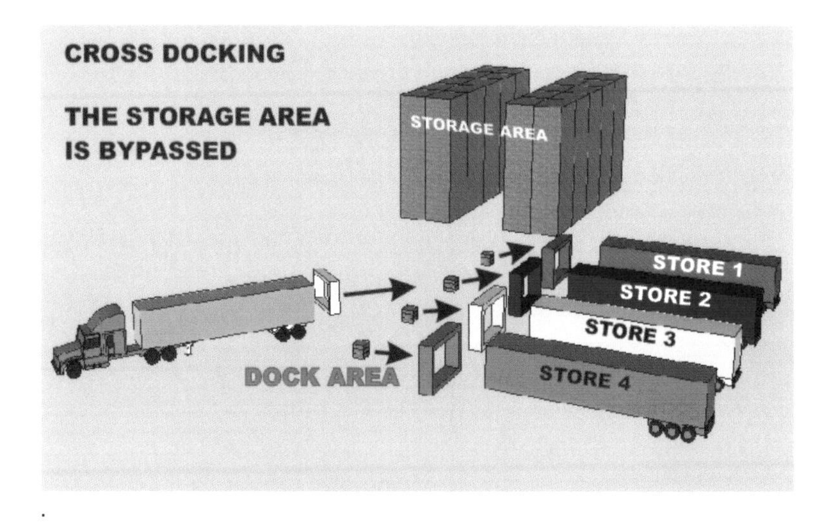

.

Tip No. 138: FORECAST YOUR SPACE REQUIREMENTS

A couple of hints here are as follows: try to express desired inventories in warehouse terms such as pallets, shelves, and whatever physical measure of inventory can be converted to space required. Also remember that increase in business income arises either from selling more of the same items or adding items. Selling more of the same items will increase your reserve storage areas while adding items will not only increase reserve storage but will also probably require additional pick spots. You will gain a valuable advantage if you can forecast future needs for space based on an actual business plan. The process of converting business forecasts into operating requirements is a complex matrix involving the relationships between the following key elements:

- Business demand which consists of the number of items or SKU's and their characteristics, the physical amount of inventory carried, customer orders and their characteristics,

- Warehouse operating resource capabilities which consist of the labor, space, equipment and control resources.

- Specific work flow elements such as receiving, putaway storage management, order selection, packing, staging and shipping that will directly affect the resource capabilities.

The process of conversion is not a formula but a logic process that must be applied to specific situations. Because every company operates slightly differently, each will have its-own priorities and questions. But the logic process works for everybody in the same way:

A. Any change in the business demand will affect the warehouse operating system and should require modification or adjustment to the resource capabilities.

B. A change can be described as a formal detailed forecast, a “what if list”, or an opportunity to test the present operating system to failure or destruction by creating your own forecast.

C. Business demand will create changes that can be identified and expressed in terms of the work flow, and the process of the product moving through it.

D. Once the work flow changes have been identified, you can then develop which warehouse operating resource capabilities will be affected, and to what degree they should be modified or adjusted.

E. Finally, one option that always should be considered is doing nothing”. This option provides the base or benchmark to which other options can be compared.

Tip No. 140: HIRE OUTSIDE SPACE WISELY

Before you hire outside space look carefully at all the suggestions offered in this book. If you absolutely must hire outside space try to make it on a short-term basis to solve a seasonal growth in inventory. Be careful with the negotiations for the space so that the relative cost of in and out storage and long-term in space storage is balanced. The third party warehouser would like to see minimum storage and maximum activity and that is where they make their money. Be very specific in the services you buy. One can buy storage space and handle material with your own employees or one can specify the third party to handle receipts, inventory taking, and shipping to your customer. The costs are vastly different and should be carefully evaluated as to which course you will take.

Tip No. 141: STORE MATERIAL OUTSIDE

If you have goods that do not require protection from the elements, consider storing them outside. This can be temporary or a permanent part of your warehouse plan. Sometimes it is cost-effective to build a simple roof shelter over the goods to protect them from direct rain damage.

Tip No. 142: MATCH STORAGE TO THE STORED ITEM

A common source of honeycomb in racks is the storage of a short pallet load in a relatively tall rack section. With the use of a proper WMS (warehouse management system), incoming material can be directed to a rack position that matches the load height of the incoming goods. This can also be done manually by walking through the warehouse, noting and rearranging short loads into short rack sections and tall loads into tall sections. Either method implies having both types of rack in the warehouse.

Tip No. 143: USE GUIDED PUT AWAY

An acceptable warehouse management system will have the ability to predetermined appropriate and precise storage locations for incoming goods. Ideally these locations will be selected for appropriate size, zone (specific area of the warehouse designated by nearness to shipping), and other factors such as security, quarantine or other factors. In this way, the worker putting loads away will work more efficiently, handling will be reduced, and loads will be placed for the best utilization of space.

Tip No. 144: CONSIDER A DIFFERENT PALLET HEIGHT

Depending on the clear height available, choose a pallet height by adding or subtracting levels of cases in order to truly utilize the available space. In other words, three tall pallets with a lot of empty space above the top is not as good as four shorter pallets that completely fill the available height.

Tip No. 145: TRACK EMPTY LOCATIONS

There are only two difficult things in warehousing. One is finding the goods to ship and the other is finding a place to put incoming goods. The next best thing to guided put away is a warehouse management program that keeps track of empty storage spots. In the absence of such a program, it is suggested that the warehouse manager walk the warehouse regularly and make a list of empty spots. When you wish to put loads away manually, there'll be no need to "cruise" the warehouse looking for empty locations.

Chapter 9: ORDER PICKING AREA

This chapter will contain a number of tips that apply specifically to smaller items. Nevertheless, remember that the principles set forth generally apply to both palletized goods and small goods on shelves or some other form of storage equipment. Since the order picking area often includes reserve storage as well, please review the tips mentioned in chapter 8, Storage Area .

Tip No. 146: CAREFULLY CHOOSE STORAGE AIDS
For example tiny or small lose pieces may be stored in shelf boxes. These in turn could be used in close show it abandons open shelving or even in carton flow rack. This would keep small parts from spreading on so I in the shelves and making the full height of an individual shelf unusable.

Tip No.147: USE FLOW RACKS
The flow rack can be either for cases or for pallets.
Pallet flow racks are often used for staging or for very high volume picking. They are comparable in cost per pallet of capacity to Automated Storage and Retrieval Systems.
Case flow racks operate on a basis of First-In-First-Out. They consist of wheeled lanes that are loaded from the back and carry stock to the front by gravity. One of their greatest advantages is concentrating a great number of picking faces in a very small bit of space.

Consider 25 items that are case presently picked from pallets in a dedicated forward pick line. Each takes up a pallet width (4 feet) and a pallet depth (4 feet). If the line was spread out it would take 25 x4 ft x 4 ft = 400 ft^2
to hold every item. An ordinary carton flow rack
could handle five across and five high lanes of cases.
A 5 foot width 7 foot height and 8 foot depth yields a space footprint of only 40 ft^2 and holds the same number of pick faces. The following tip suggests placing pallet rack above

the flow rack to utilize the height above. Incidentally, the use of the flow rack reduces the pick travel for the 25 items from 100 feet to 5 feet. This is reflected in faster pick times.

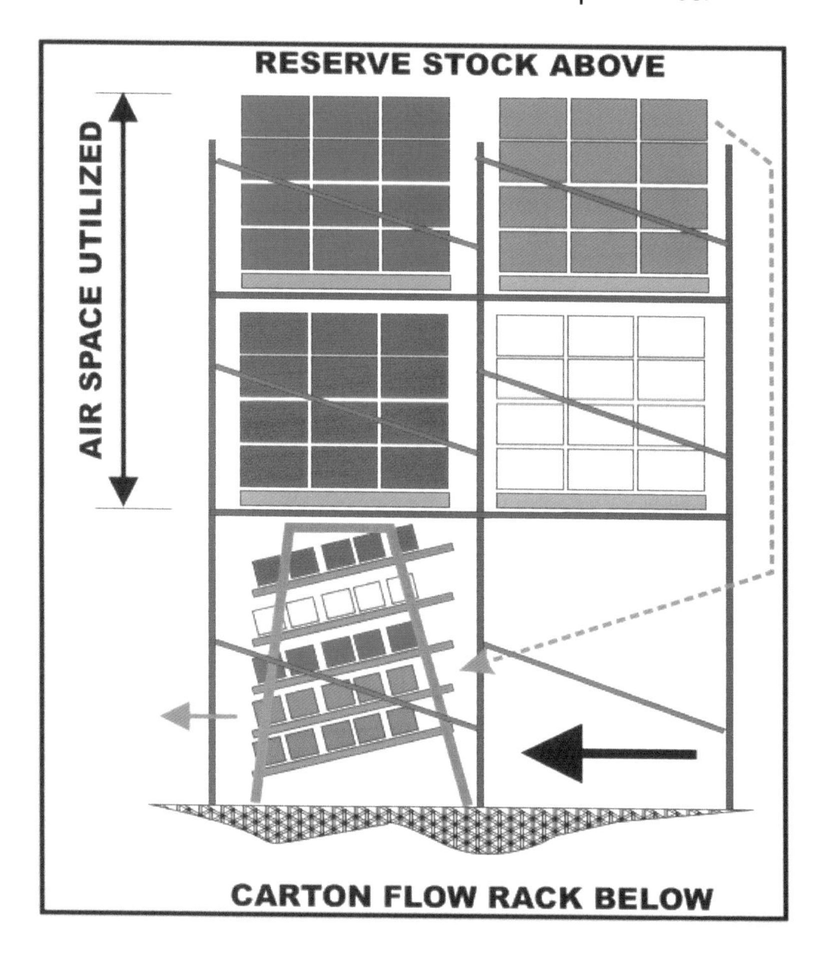

Tip No. 148: USE PALLET RACKS ABOVE FLOW RACK

The use of pallet rack above the carton flow rack makes good use of the air space above the flow rack. As shown above, it can be used to accommodate the appropriate reserve stock.

Tip No. 149: USE PICK TRUCKS

Sooner or later we all realize the wastefulness of open space soaring above our 10 foot high small parts shelves. The manner in which we utilize this cube depends on the type of small parts picking as well as our budget. One approach is to stack shelving one upon another to build up the height to match the available clear height. The problem now becomes one of reaching the shelves which are now well above the floor. One way to do so is to utilize the "stock picker truck". See Chapter 2 Handling Equipment, Aisles, Storage Height.

The "Stock Picker" is similar to an ordinary fork truck except for one vital difference; the operator platform rises up with the forks. This allows the pick operator to reach any level throughout the system height. Since it is usually a narrow aisle system, the operator can reach both sides of the aisle. An appropriate safety harness prevents accidents when reaching out from the sides of the machine

Tip No. 150: ADD A MEZZANINE

Another approach to utilizing the height above your small parts shelves is to literally stack one or more sets of shelves atop the existing ones. If the originals were properly bought, this is no more than literally placing the second set on top and using special clips to hold them together and to serve as a holder for a drop-in flooring section of expanded or lattice work metal. Suddenly we have doubled or even tripled the storage area by using the storage cube. That is the good news but what is the possible bad news? It may be the fact that now we have a set or more of stairs to climb in order to reach all of the goods in the system. One way to minimize this is to apply "Pareto's Rule". Pareto's rule comes from the work of Vilfredo Pareto, a 19th century Italian mathematician, (working in Switzerland) who discovered that 80% of the wealth of Switzerland was concentrated in just 20% of its citizens. The significance of this is that the same rule applies without fail to warehousing situations!

For example, 80% of the action in your small parts area will occur in only 20% of the items! Our solution then is to identify the fast movers and make sure that they are on the first floor of our new mezzanine system. Surely there will be some trips up the stairs but the vast majority of the calls will be on the first floor in the active group. It should go without saying that a decent addressing system is a must to implement such a plan.

Tip No. 151: USE HALF PALLETS FOR SLOW MOVERS

Handle slow movers on half pallets, and reduce space requirements both in storage and at the pick face. Assume that standard pallet racks are used. Front-to-back members can be introduced into the racking so that the half pallets or skids are supported. Then, the following method is used:

1) An incoming half skid is delivered to the slot where it is to be stored;

2) the fresh load is set down on the floor in the aisle;

3) the fork truck operator backs away, turns in, and picks up the last half-pallet and sets it down on the floor in front of the

fresh pallet -- by thinking ahead even further, the fresh pallet is picked up with the old stock in the proper position;

4) the two half-pallets are returned to the storage location. This system could work in almost any conventional warehouse that presently uses the 48-by 40-inch pallet rack bays with 93-inch openings.

Tip No. 152: SIZE PICKING SLOTS APPROPRIATELY

Consider the speed at which picking slots are emptied when sizing them. Also consider how often you wish to replenish these slots. If all the slots are the same size for both fast-moving and slow-moving items, you will find the slots for the slow-moving items are tied up for a very long time. It may be a good trade-off to make smaller slots and more frequent replenishments in order to save space on the pick line.

Tip No. 153: NARROW THE REPLENISHMENT AISLES

By narrowing the aisles used for access and loading from the other side, when using carton flow rack you can reduce the space needed to replenish the inventory. Compared to a conventional system, one aisle can be used for picking and another aisle for replenishment and thus beat the conventional system in terms of real storage efficiency. Another hidden benefit is automatic rotation -- FIFO is going to be virtually automatic.

Tip No. 154: CAREFULLY CHOOSE STORAGE AIDS

The use of storage aids to consolidate and hold is a space effective way to reduce the space used . For example tiny or small loose pieces may be stored in shelf boxes. These in turn could be used in open shelving or even in carton flow rack. This would keep small parts from spreading out in the shelves and causing the full height of an individual shelf to be unusable.

TIP No. 155: SEPARATE SLOW MOVERS
By using high density storage rack systems and mezzanines for order selection of slow movers, you can arrange the fast movers in easily accessed selection aisles and at the same time reduce the width of the aisles to the minimum.

Tip No. 156: USE FORWARD SEPARATE PICK SPOTS

Many organizations pick orders from floor stacks of pallets. Those cases required for a mixed pallet are picked from the pallet stack that it closest to the aisle.The following steps must be fulfilled in order to pick a full pallet.

A. Pick up the partial pallet in the first row and set it down in the aisle.
B. Pick out the full pallet wanted and set it down in the aisle.
C. Turn and pick up the partial pallet and replace it in the first row of the stack.
D. Pick up the full pallet in the aisle and transport it to be shipping floor.

Changing this method to the use of separate pick lines for cases and pallets will not only gain picking efficiency but will also gain space by using a narrower aisle in the case picking section as opposed to a wide aisle where pallets are picked.

Tip No. 157: CONSIDER ADDITIONAL PALLET SIZES

Although we should avoid a ragtag collection of sizes that lead to difficult handling and inefficient cube utilization, consider the advantages of changing certain items on the pick line to a pallet size that is a multiple of the main size.

For example, consider a case pick line on 48" wide pallets. Let's assume that the business has grown by adding items and that we suddenly find that we do not have enough pallet positions to have only one item per pick pallet. Mixing items on a pallet is an invitation to error but In many cases the answer may lie in planning a smaller load size and putting three pallet widths in the space now occupied by two, e.g., 3 x 32" = 2 x 48".

Additional changes might lead to a reduction of load height for certain items in order to add a load beam, stacking the items two high and three wide where formerly we had only two positions. All six of the new pick positions are reachable from the floor, and we have effectively tripled the number of pick positions. It should go without saying that a careful study of inventory levels and cubage is the key to making this change effectively. Replenishment costs are also part of the tradeoff equation and should be considered.

Tip No. 158: CHOOSE A NARROW PICK AISLE

Within reason, the narrow aisle principle applies to small-parts storage as well as it does to unit-load storage systems. Many planners specify too wide an aisle for simple shelf storage; this turns a single-pass pick path with easily reached cross-aisle picks into a path that goes up one side and back the other, thereby doubling the travel distance. The same reasoning applies to very tall stockpicker truck systems. Often the one-way narrow aisle is the right choice in a controlled storage system. And it saves space!

Tip No. 159: **CUBE UTILIZATION**

Adherence to the principle of cube maximization is perhaps even more critical in the management of small parts storage and order selection than in unit-load storage, since a relatively small amount of honeycomb (wasted space) can represent a very large percentage of the storage space designated for small parts. In a sophisticated and expensive system, every cubic inch is valuable. The way to accomplish good cube utilization within the storage system is to plan the storage compartments in a way that corresponds to the requirements of the items to be stored. The use of the cubic measurement of the stored inventory is a valuable tool for accomplishing this.

Tip No. 160: **BALANCE ACCESSIBILITY WITH SPACE EFFICIENCY**

An exaggerated example of high space utilization with poor accessibility might be for filling a storage drawer to the very top with ten different kinds of electrical resistors. The available space in the drawer is certainly completely used, but the access time to find a particular resistor will be ridiculous. A better solution is to use bins within the drawer to separate the parts and thus improve access time even though sacrificing part of the available space. Trade-offs are as important in small parts storage and picking as in any other warehouse situation.

Tip No. 161: AVOID VERY DEEP SHELVING

It would appear that the use of extra depth shelving would favorably increase the ratio of aisle to available storage space. In theory it looks good, but the fact is that a deep shelf provides a hiding place for stray inventory; things invariably end up behind some other item in a semi-invisible spot. Yet another problem arises when a deep shelf is set so close to the one above it that it is extremely difficult to reach in for the inventory at the back.

Tip No. 162: ADJUST SHELF HEIGHTS TO THE GOODS

Look at your shelf storage and resolve to adjust the shelf heights to match the goods stored. The same reasoning applies to drawer dividers in storage drawers.
If you buy new shelving, make sure the shelf heights are easily adjustable. Always match the container to the inventory stored. Note that this is the very same principle applied to pallet storage in racks.

Tip No. 163: EVALUATE CAROUSELS

If you have ever watched the clerk at the dry cleaners retrieve your suit from a hanging endless chain conveyor, you have observed the basic model of the warehouse carousel. Shelf sections are hung or floor mounted on an endless track (see below) and the push of a button activates the carousel. Indexing is available so that a specific section can be made to stop in front of the operator. This is an example of bringing goods to the picker rather than picker to the goods. Carousels may also be computer controlled, mezzanine mounted, and automatically loaded and unloaded for an automated system. Carousels have even been linked together in a high-rise configuration under complete computer control. The horizontal carousel saves space because multiple units can be mounted side by side with little or no aisle.

HORIZONTAL CAROUSEL

Tip No. 164: SLOT FLOW RACK BY SIZE

The use of space is also influenced by the layout strategy for the pick line. This is the place to recognize that it is a constant compromise between labor efficiency and space utilization. For example if we were to layout smaller goods in shelves, we would certainly attempt to place the goods with consideration of the expected activity. This is all well and good but what if some of the goods are taller and some shorter. When a tall and a short are placed side-by-side, the height of the shelves must accommodate the taller item. This leads to a lost space or honeycomb above the shorter item. In this situation it might best to sub optimize the situation by storing by item height within an area generally reserved for faster item. See illustration below.

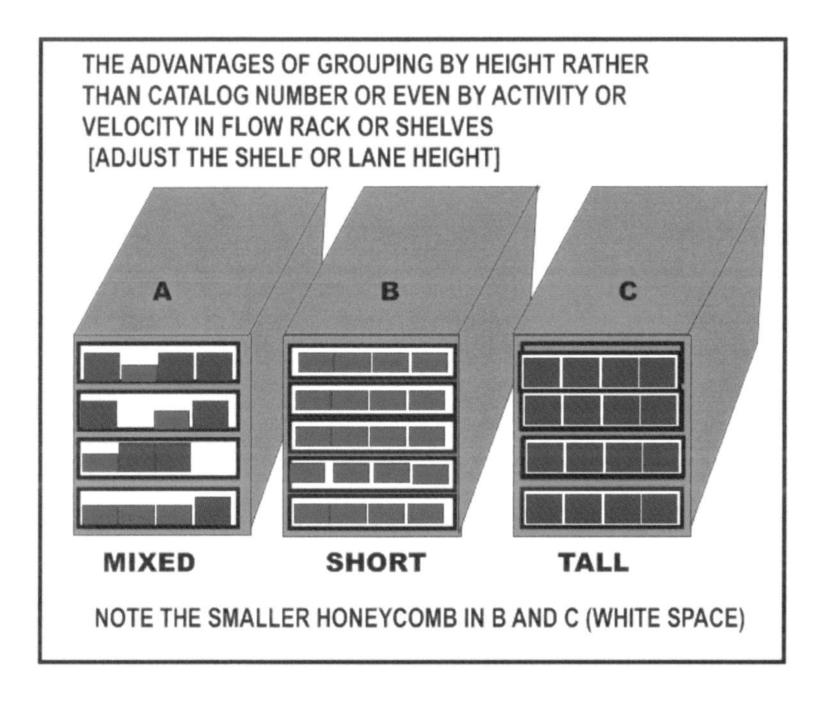

As shown in the illustration, varying package heights can make the flow rack cube utilization very low. Another method keeps items in numerical order for simplicity. In both situations, it may be a greater advantage from the standpoint of available pick slots, to group the stock by size. The illustration shows an extreme example but the principle holds. When this is implemented, not only are more slots available but the cube utilization is increased and the pick travel distance reduced.

Tip No. 165: PICK SMALL PARTS FROM A MODULAR DRAWER SYSTEM

(See Chapter 2, Tip 29) Although useful for almost any picking system for small parts, this is most often used in high value parts and tool rooms. Essentially, the system is composed of drawer units that pull out to expose the entire storage area, and extremely smooth drawer slides that are not only easy to open and close, but can carry up to 400 pounds per drawer. These are combined with a series of accessories that allow infinite division of the storage area. What makes it special is the careful thought that went into the design of a system that allows complete interchangeability and variety of storage. One of the benefits of organized storage is that by matching the storage slot to the items stored, honeycomb is kept to the absolute minimum.

While in the past, we have demonstrated the evils of honeycomb with examples of short pallets on tall rack openings and other large load examples, the basic principle applies equally to such things as small parts placed on oversized shelves. In fact it is not overly optimistic to gain space reductions of 80% in the change-over from shelving to Modular Drawer Systems. Where floor space is very tight or extremely expensive, the MDS can be justified on the basis of reducing the space needed to store a given quantity of parts.

Chapter 10: PACKING AND CHECKING AREA

Rarely is very much thought given to either the packing area or the order checking area insofar as efficiency of procedure or space utilization. A little attention can go a long way in these areas.

Tip No. 166: IMPROVE PICKING ACCURACY

An improvement in the accuracy of order selection can and should lead to a reduction in order checking. This in turn can move material through the checking area more swiftly. With a smaller backlog of goods to check, the checking area can be reduced in size thus saving space. With a powerful WMS and bar code Radio Frequency terminals it is actually possible to eliminate checking completely. It follows that the former checking area can be put to other uses.

Tip No. 167: START STATISTICAL ORDER CHECKING

A similar space saving (and labor saving) can come from the institution of sample order checks and true statistical order checking rather than 100 percent item and quantity check of orders.

Tip No. 168: KNOW YOUR ERROR RATE

If you are not making errors, cut back on the amount of checking. When an acceptable rate is achieved, it may be possible to reduce the size of the order checking area.

Tip No. 169: STORE SUPPLIES OVERHEAD IN THE PACKING AREA

The 'air space" above the packing tables or pack lines can be utilized to hold cases and other supplies within easy reach. This not only saves space but increases productivity.

Tip No. 170: STORE RESERVE SUPPLIES IN THE LEAST ACTIVE RACK AREAS

A reserve inventory of supplies should be stored in rack positions that are not suited for regular pallet loads. This will release properly sized spots for active storage.

Tip No. 171: PROVIDE AN OVERHEAD TAKE AWAY CONVEYOR ABOVE THE PACK LINES AND TABLES

The overhead conveyor can take the empty cartons and excess "trash" that is generated as part of a repack operation. Without this solution, trash and multiple empty cartons will collect on the floor and take up valuable pack and maneuvering space.

Tip No. 172: PROVIDE OVERHEAD POWER DROPS TO THE PACK STATIONS

Packing stations need power tape dispensers and other electrical tools to efficiently pack for shipment. Extension cords on the floor are trip hazards and also make the floor unavailable for tempoprary staging and storage.

Chapter 11: SHIPPING AREA

Expect some reiteration of principles in this section because most of the tips for receiving, apply equally to Shipping.

Tip No. 173: **TIME YOUR SHIPPING**

Save space in the shipping area by carefully prioritizing each order for specific time periods. In this way material will stay in the staging area for a minimal amount of time and therefore minimal space.

Tip No. 174: **UTILIZE THE CUBE**

You will see this one repeated in every warehouse functional area. Simply stated, it is the key to finding more storage space. It applies as well to shipping as to receiving

Tip No.175: **ADD RACKS FOR STAGING**

Where the problem of insufficient outgoing staging exists, use racks to drastically improve available staging. Balance the extra effort of lifting staged orders into the upper sections of the rack with the advantages of enough staging to plan shipments efficiently.

Tip No. 176: **COMBINE RECEIVING AND SHIPPING**

It is true that in most facilities, shipping and receiving activities peak at different times. Receiving is usually a morning exercise with shipping peaking in the afternoon. If the dock locations allow it, one staging area could serve both functions. One combined office would also serve both activities. An added benefit is that fewer managers or lead people are needed as well as getting better utilization of the handling equipment

Tip No. 177: DETERMINE THE SHIPPING STAGING AREA

The space in the shipping area is a function of the area needed to hold one day's shipping. Calculate the cube of a days shipping and convert that cube to physical units (pallets, cases, etc.) Do not overstate the needs because if too much staging area is provided, be sure that it will be filled and orders will be delayed in shipment.

Tip No. 178: SCHEDULE OUTGOING TRUCKS

By spreading the arrival of trucks to pick up and load, it is possible to reduce overload and jams at the docks and the loading area. It is a totally realistic concept that trucks can be scheduled at **your** convenience.

Tip No. 179: FILL INTERIOR TRUCK DOCKS

Another source of additional space can be found in those facilities that have inside truck docks. Filling in those docks and then using an outside truck door with a weather seal and canopy can often provide enough space to avoid a move to a larger warehouse. Most inside docks are about 15 feet wide and 65 feet long. The area to be picked up is 975 square feet, enough for up to 148 additional loads, depending on the storage/staging system used.

Tip No. 180: PLACE RACKS ABOVE THE DOCK DOORS

Racks above the dock doors can hold empty pallets or other supplies where they are needed and thereby free up prime storage or staging areas for out going shipments.

Tip No. 181: LAND USE

Zoning restrictions setbacks and rights of way can often be utilized as a place for temporary truck parking while waiting to pick up an outgoing shipment.

Tip No. 182: **STORE EMPTY PALLETS IN BAD RACK POSITIONS**

A "bad" rack position occurs when the clear height of the building is such that the top rack shelf cannot accommodate the height of a full pallet load. A short stack of empty pallets is a good way to utilize that space.

Tip No. 183: **ASSIGN ADDRESSES TO STAGING SPOTS.**

This will allow you to track shipments and their status more easily. In many cases it will empty the staging area more rapidly to load scheduled trucks. This works best if you have a real-time computer system, because the staged inventory status is constantly available.

Tip No. 184: **OUTLINE AISLES AND STAGING**

Control aisles and loading areas by marking the floor with a line marker or tape. The tendency to leave goods in the aisles or loading areas is reduced when they are clearly delineated

Tip No. 185: **ANALYZE YOUR STAGING AREA**

Ask yourself the following questions:
Is the staging area completely filled? If so, it may be too small. Is the staging area relatively empty? It may be too large and lead to excessive cycle time waiting to ship. If a staging area is provided to hold two days shipments, you may be sure that some material **will stay there as long as**

two days. When only one day's staging is provided, efforts will be made to ship goods before the end of the day. This is a good thing!

Tip No. 186: USE CROSS DOCKING

If incoming material is needed for distribution to a number of predetermined destinations, it makes good sense to ship it out immediately without putting it away. Taken directly from receiving to the outbound docks. This will reduce your order cycle time and clear any staging areas efficiently.

Tip No. 187: CLEAN UP THE WAREHOUSE.

The cleanliness of a warehouse shipping area is very important. In dirty warehouses dust will settle on cases of merchandise and on freshly packed shipments. A dirty warehouse provides unpleasant working conditions and will lower employee morale.

Tip No. 188: SCHEDULE FOR DIRECT TRUCK LOADING

Outgoing staging may be reduced by organized shipping procedures that allow a greater proportion of direct loading rather than pre-staging. This is a matter of balancing truck arrivals with order selection. Basically, the truck waits for you, not the other way around. In spite of the threat of demurrage charges, trucking is very competitive and you can negotiate by placing your business where the cooperation is greatest.

Chapter 12: PLANNING A NEW FACILITY

TIP No. 189: EVALUATE TRADE-OFFS

Nothing is ever perfect in life so why should we expect it to be so in our warehousing and distribution operations. An example will make our point. Consider a particular enclosed space or room. (See illustrations below) what is our best bet for ideal space utilization? Obviously we could fill the room

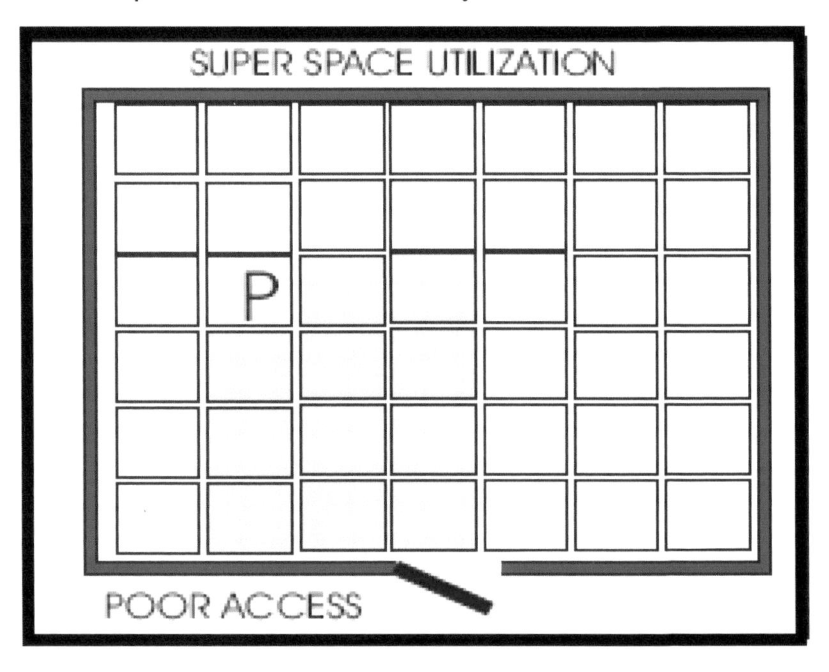

to the walls and all the way to the ceiling with pallets. We could shove the last ones in and slam the door. We have utilized the space almost 100 percent! BUT, what if we want to retrieve pallet **"P"**? All of the loads in front must be moved in order to access the needed one. Our handling productivity is intolerable. Let's go to the other extreme. (See below). We show just a few pallets placed in the room so as to allow full fork truck access from all sides.

Now we have solved the access problem but are in terrible shape for space utilization. The answer of course is a compromise or an optimum solution with both **reasonable** access and **reasonable** utilization of space.. Please note that these are but two of the factors that influence design. Nevertheless, they must always be in the back of your mind when designing new warehouses.

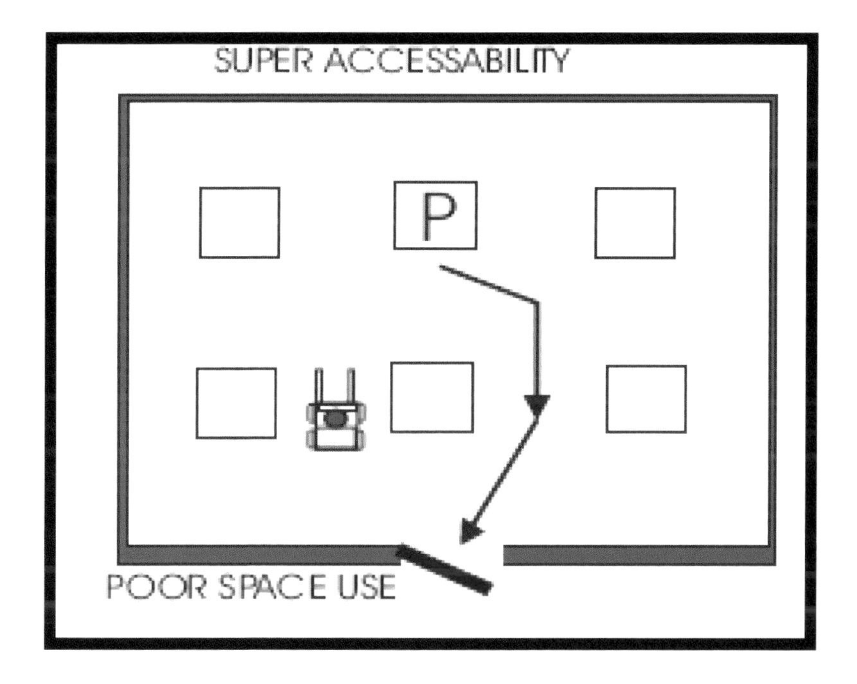

TIP No. 190: ESTABLISH PHYSICAL REQUIREMENTS AND TIE THEM TO A BUSINESS PLAN

SHRINK OR GROW?

In either a growth situation or one of a reduction in sales or inventory, it is of prime importance to avoid guess-work and surprises. Both situations can increase costs and possibly create bottlenecks that can strangle a business. In the case of a reduction of activity, it can be useful to know just how much space will be needed under the reduced conditions. It may be possible to combine locations of multiple facilities or to reduce the use of public warehouses. It may even be possible to reduce space usage enough to allow a portion of your warehouse to be leased to a non-competitive business. At any rate, many things are possible when the real facts are known and when they are expressed in real and physical terms that can translate directly to the layout of an efficient warehouse.

In the midst of what some have chosen to call an economic adjustment and others refer to as a recession, some businesses continue to grow and prosper. Many are faced with problems of rapid growth, necessitating physical expansion plans for warehousing, and, while developing analytical techniques to accomplish the planning, we realized that this approach works the other way as well. That is to say, the method that relates increased sales to increased physical requirements also yields a rational approach to decreased physical requirements when sales decline. So if you are in a sales decline and have multiple warehouses that might be combined, the techniques described will predict the space needed. This approach of expressing sales patterns in terms of warehouse space can help implement profitable growth. If business is shrinking it will be a way of pulling in your horns in a realistic and efficient way.

TIP No. 190: PREDICT THE SIZE OF THE WAREHOUSE

The approach used by management to predict an appropriate warehouse size and capacity has not changed very much since Biblical times. This is when Joseph, that ancient consultant, translated Pharaoh's dreams of "seven cows and seven stalks of grain" into an action plan calling for warehousing 20% of the Egyptian crop. *"And let them lay up corn under the hand of Pharaoh, and let them keep food in the cities." (Gen. 41:35)*

We sometimes interpret the dreams of our management with much the same zeal but without the divine guidance available to Joseph.

TIP No. 191: WAREHOUSE REQUIREMENTS ARE NOT PROPORTIONAL TO SALES CHANGES

It is a fact that warehouse needs do not increase proportionally to growth. A 50% increase in overall business does not correspond to a 50% increase in warehouse area. This is, of course, primarily due to differences in mix of the products in storage.

PRODUCT MIX

Some items live in the warehouse a long time and some a very short time (low number of turns versus a high number of turns). Low turn items take a relatively large amount of space for a given amount of sales. Conversely, fast turning items take a relatively small space for a given amount of sales. If the increased sales forecast is heavy with low turn items, the space requirements will go up faster than if high turn items pre-dominate.

STORAGE CONFIGURATION

The second factor influencing space planning predictions is the fact that varying storage methods or configurations yield different space needs in square feet. Unfortunately, most managements still think in terms of square feet rather than cubic feet. The result is that a typical plan not only assumes problem by expressing the growth in terms of the same low

headroom area presently in use. Since it is a sad fact that the warehouse people only hear of a new warehouse expansion after it is planned by the architect and the foundations are down, there is little to be done but shoehorn a layout into the new facility. This tends to perpetuate the methods of storage existing in the old warehouse.

WHAT IS REALLY NEEDED

The ideal way of forecasting and planning a new warehouse (or reducing and consolidating multiple ware-houses) is to express the needs in warehousing units, i.e., loads, shelves, bins and the like. When this is done, the numbers can be expressed on a plan or layout displaying specific storage configurations, heights and column spacings.

The importance of warehouse units is that they represent storage and movement much better that the usual parameters, such as dollars of inventory, or even units of product. How much space is needed, for example, for $10,000 worth of widgets? Even knowing that this represents 1000 widgets still leaves us wondering if a widget more closely resembles a transistor or a bulldozer. How much more useful to know that the widget inventory is 24 pallet loads 48" x 40" x 60" high. Now we can plan for specific racks or floor stacks and evaluates the storage methods in economic terms.

Building space too can be estimated at a particular height and accurate dollar figures obtained. Furthermore, if these realistic plans and figures for the warehouse can be tied to a variable sales forecast broken down by product mix, we now have a flexible, useful and realistic planning tool. This can also be used as an operational aid once the program or manual computation technique is in place. Later in this article we will discuss how the results can be used to avoid warehousing surprises and to evaluate selling margins from the standpoint of warehousing cost.

HOW TO PLAN IN WAREHOUSING UNITS

As in so many parts of warehousing technique, here too the vital ingredient is cube. The information that can be imparted by this simple bit of data is invaluable, and if it is not now part of your data base, take the necessary steps to include it. To plan in warehousing units, the following information is necessary:

A. Sales or movement by units of issue

B. Actual or desired turns by item, or

C. Desired inventory in terms of unit movement per time period (e.g., an inventory equal to six weeks of sales, or any given period, depending on reorder time, restock levels or delivery times)

D. The cube of a sales or issue unit

E. A realistic record of movement over a representative time

F. Ranking of items by sales volume and by cubic feet of inventory

G. An idea of the general methods of storage that might be used, e.g., pallets of 48" x 40" x 60", or standard shelves 31" wide x 21" deep x 11" high, or bins 21" deep x 41" wide x 24" high.

All of these storage methods have a definite cubic capacity: pallet = 66.67 cu ft., shelf = 6.00 cu ft., bin = 16.00 cu ft.

If we list all items along with an average week's activity, multiply each item by the number of weeks in the desired inventory to get the desired inventory in units and then multiply the inventory by the cubic feet per unit, the result will be an inventory in terms of cubic feet. If the items are listed from largest cubic inventory to smallest, we can then divide the cubic feet by the cubic
feet of the desired storage method, and the result will be an inventory expressed in warehousing units. A consideration of size will indicate the probable best storage method.
For example, a total inventory in hundreds of cubic feet will tend to indicate unit load storage and, if large enough, may also imply the desirability of floor stacks. An inventory of 30 to 50 cubic feet translates to less than a full pallet but indicates the possibility of case picking from a total inventory of one pallet slot.

The points at which the storage methods change are somewhat flexible and depend on an understanding of the products and how they are issued and stored. One way to test is to divide all inventories by 66.67 cubic feet (a pallet cube) as a first cut and see how they look. If the numbers rapidly change to small four place decimals, we will know that we must use a different and smaller storage method for the items.

Some break points that have been used successfully in past projects are:

A. Greater than one pallet to one-half pallet (greater than 67 cu ft down to 33 cu ft)....on pallets.

B. 33 cu ft down to 6 cu ft—multiple shelves or bins or case flow rack.

C. 6 cu ft down to 3 cu ft—full shelf.

D. 3 cu ft down to 1.5 cu ft—half shelves.

E. 1.5 cu ft down to .75 cu ft—quarter shelves.

If smaller breaks are used, one can predict drawer sections.

Several facts should be kept in mind when analyzing the output. First, a fractional pallet takes the space of a full pallet; second, a large number of pallets implies a pick pallet plus the rest in reserve (unless the pick unit is a full pallet); and third, it is always necessary to round up the number of units until the warehousing unit changes, as when going from partial pallets to shelves.

ESTABLISHING INVENTORY LEVELS

The approach just described and is based on the ideal situation of already having a data base that includes cube, unit sales figures and a clearly stated inventory policy based on weeks of usage. Sad to say, this is not always the case. Typically, cube is not known and the inventory levels are not understood in terms of usage.

The first step, therefore, is to uncover the desired inventory levels through discussion with management or the director of purchasing. This dialogue can serve another purpose as well, as it will start to acquaint those who buy with the effects that this buying will have on the warehouse. This is in addition to the beneficial effects of thinking in terms of warehouse mix. The ideal outcome of the exercise is to generate a sales plan in terms of unit inventory which may be expressed in warehouse units and subsequently be translated into actual physical space and layout. A review of the final results may even cause management to modify its plans because of the effect on the warehouse. Although it is a fact that management considers the dollar value of inventory, rarely does it consider the physical impact of inventory on the distribution system.

GETTING THE CUBE

Somewhere along the line, a physical measurement is necessary. The warehouse personnel must measure or the information will be available from suppliers or the package engineering department of your company. If measurement is the approach to be taken, use shortcuts as much as possible. For example, below a certain size very small items, such as transistors or small screws, will not greatly affect the warehouse size. Find out where the level of accuracy lies and actually construct a sample of the object which is the smallest cube to be considered. All things smaller will then be entered at that figure. As an example, suppose that anything below .001 cu ft will be considered .001. A model 1" x 1" x 1.5" would be used as a comparison standard, thus simplifying or avoiding many laborious hand measurements. In addition, if the diversity of the inventory warrants it, rent or buy a **CUBISCAN®**. This is a device that automatically records dimensions and cube and enters the information into a database.

THE 80/20 APPROACH

Even with the straightforward approach outlined above, the task is a large one and may seem almost insurmountable if the product line runs into the thousands. There is a way to approximate the results while significantly reducing the effort involved. Thank Dr. Vilfredo Pareto for this, because we are referring to Pareto's Rule or the 80/20 Rule, as it is frequently called. Simply stated, it says that 80% of the activity (and therefore 80% of the inventory)is concentrated in about 20% of the items. If there is a way to identify the 20% of the items that are most active, work can be done with that group with reasonable confidence that the major portion of the warehouse activity has been covered. When a universe of 10,000 items is considered, it will be cut to only about 2000 items requiring laborious manipulation and measurement.

In order to reduce the amount of measurement, it is necessary to make a simplifying assumption: that unit activity is an indication of inventory size—not strictly true, but usable as a first cut to establish a list of 20% of the fastest items. The entire list of items should be run with a cumulative total of units moved, so that it is possible to see where the 80/20 break occurs. The items above this point are the ones to be used for analysis. shows the items with the inventory levels in weeks inserted. When this is multiplied out, we will have a total inventory in units for each item on the sheet. If cube is entered for this short list of items, the calculations can be continued to yield cubic feet of inventory for each of the items. This, in turn, is divided by the appropriate cube of a pallet load (e.g., 66.67 cu ft) to yield inventory loads. These should be arrayed from the largest number to the smallest. In this way the break points can be established for change overs from one sort of storage system to another.

ESTIMATING THE REST

Many of them will be fractional shelf items because of their low inventory. The inventory is expected to be low because of its dependence on activity. Certain items, of course, because of their large cubic size, will be on pallets, and others will be larger because of special buying conditions which out-weigh their inactivity. The first run should be scanned for such items by an individual familiar with the business.

In summary then, the extrapolation of the rest of the warehouse should be done by assuming a typical warehouse unit for each item and multiplying by the remaining number of items to give physical reality to the 80%.

PRESENTING THE RESULTS

A list can be prepared showing the distribution of items for each type of storage. Knowing the number of items and the inventory of each in warehouse terms, plus the fact that whenever the inventory is not wholly contained in one warehouse unit there will be a pick spot and a reserve, allows us to plan not only storage but pick lines as well.

To change the planned storage method (e.g., to go to a different pallet size), it is only necessary to multiply by the old pallet cube and divide by the new. This changes the figure to loads of the new size. This can be useful when contemplating a change to a smaller pallet in order to open up more floor pick spots. Once more, remember that a partial pallet takes up the same amount of space as a full one.

USING THE INFORMATION

Now that we are in possession of a physical interpretation of the warehouse derived from activity, we can see what will happen as that activity is changed. If management has predicted growth in certain segments of our mix, we can recalculate the physical cube of the affected inventories and assess its effect on the planned warehouse. If management decides to lower inventories by buying a shorter supply, we can change the calculations accordingly and see a fairly realistic picture of the effect.
It is completely possible that a single product line can so affect a warehouse as to make it a bad decision to expand that line at this time! This could be the proverbial straw that breaks the warehouse's back. The method described allows us to predict this in advance and give input to management that may affect its decisions.

A variation of the above reasoning allows the evaluation of a given warehouse in terms of how long it will serve the sales forecast without a move or a major expansion. A relayout and upgrade of equipment may simultaneously be indicated and justified by the same exercise in which it is shown that the warehouse in its present state will not meet the forecast. A timely switch to narrow aisle or higher cubic storage may give enough additional storage and pick slots to fill the proposed sales plan.

OPERATIONAL USES

The use of this prediction device can be used on a daily operational basis as a short range planning tool. If the purchasing department regularly advises the warehouse staff of new products or special purchases planned, along with their item cube and expected rate of sales or movement, the warehouse operators can plan ahead for a known quantity of loads, shelves or bins to be stored in specific warehouse locations appropriate to the expected activity. If the calculated bulk of the item is too much for the warehouse, temporary outside warehousing can be negotiated without a sense of desperation.

In some cases, communication on the subject of new stock can lead to a beneficial modification of the buying plan before the commitment is made. The ability to avoid surprises can be one of the biggest advantages of this system that predicts warehouse needs in advance.

PRODUCT MARGINS

One of the still neglected areas in warehousing and distribution today is the assignment of warehousing costs to specific products or product lines. The outlined technique can be applied to an individual product line to give a realistic measure of the space needs. This is a start towards identifying the other distribution costs of an item. It is probably the most difficult to assess under present conditions.

HOW DOES THIS AFFECT PLANNING A NEW FACILITY?

WAREHOUSE DESIGN CRITERIA

WAREHOUSE UNIT : UNIT LOAD PALLETS

NO. OF ITEMS	NO. OF PICK SLOTS	NO. OF RESERVE SLOTS
175	175	45

WAREHOUSE UNIT : FULL SHELVES

NO. OF ITEMS	NO. OF PICK SLOTS	RESERVE SHELVES	TOTAL SHELVES
400	400	250	650

WAREHOUSE UNIT : HALF SHELVES

NO. OF ITEMS	NO. OF PICK SLOTS	RESERVE SHELVES	TOTAL SHELVES
350	350	0	175

WAREHOUSE UNIT : QUARTER SHELVES

NO. OF ITEMS	NO. OF PICK SLOTS	RESERVE SHELVES	TOTAL SHELVES
640	640	0	160

NOTES: Divide total shelves by 7 to convert to shelf sections

Part of "A" and "B" and all of "C" and "D", total inventory is on one load, shelf or fractional shelf.

THE DESIGN CRITERIA

Consider the design criteria as presented on the previous page. It is really a complete specification for the warehouse requirement expressed in warehouse units. This should be an outcome of the previous methods outlined.

Now that you are equipped with a set of warehouse requirements in physical warehouse terms, it is a fairly straightforward task to lay out a new facility or to relayout parts of an existing facility. Given the storage units and the numbers, you can store "X" Pallet loads in any of the many configurations and heights outlined in previous chapters. Similarly, a known requirement for shelves can be high or low storage, tall shelves or mezzanines as efficiency and justification dictates. The specifications can even be used to plan and layout a totally automated Automatic Storage and Retrieval System (AS/RS) warehouse if so desired.

APPENDIX

HOW TO DESIGN EFFECTIVE PALLET PATTERNS

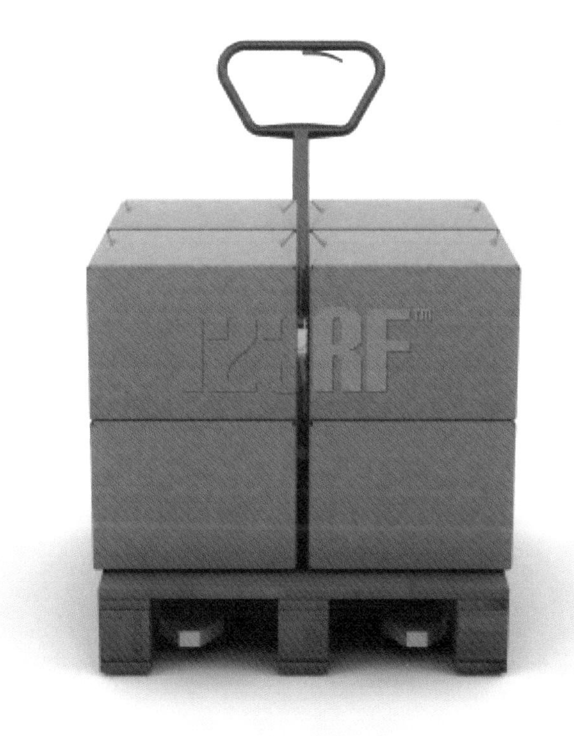

PALLET PATTERNS LOGIC AND THEORY

There is an old cliché that says,"Those that can't do, teach." A modern variant might be, "Those who no longer do, manage." Neither of these indictments should be taken too literally; their essence is found in the realization that sometimes, in the practice of education or management, we tend to forget what it's like out on the floor. Consultants are also prone to this lapse; there are many detailed tasks that are more satisfying to recommend than to perform. In the slightly removed fields of teaching, managing or consulting it becomes possible to forget the realities of day-to-day activity.

An example of a subject that has suffered from this effect is the entire area of palletization and pallet patterns. We take it for granted, and the days when it was considered worthy of hard thought are long past. How long has it been since you put together a unit load of incoming or outbound goods? We are talking not about issuing instructions to a subordinate, but about physically picking up cases and deciding on their position on the pallet.

These thoughts are dedicated to a review not only of the basics of palletization and pallet patterns, but also to a discussion of the logic of and rationale for these subjects. We feel that this will help those who have not been exposed to these ideas, as well as refocusing the attention of those who have not had to think about them for years. We hope that it will cause you to review your own pallet pattern procedures and perhaps to provide the tools of uniformity and the guidelines that are sorely needed by the people out on the floor.

PALLET PATTERNS

In most warehouses the major portion of the available space, and therefore the major part of the warehouse cost, is taken by palletized storage of unit loads. The unit load is therefore the most important common denominator in the warehouse.
The storage and movement of the unit load receive a large share of attention in seminars, handling magazines, and the pronouncements of consultants. It is understandable to feel that the last word about pallets has been said long ago. Moreover, some warehousemen feel that the tedious job of working out the most efficient pallet pattern is hardly worth the trouble it takes.
The truth is, however, that both of these ideas are dead wrong! Since unit loads usually represent high volume items, and since large savings are to be found in improving high volume operations, improvements to unit load design can yield great dividends.

UTILIZE THE CUBE

"Bottom area efficiency" refers to the degree of utilization of the load base or pallet surface. Obviously, the more packages that can be loaded onto a pallet, the greater the storage efficiency. It takes a fixed amount of space to store a pallet, whether it is empty or full. The result is that bottom area efficiency does influence storage effectiveness and therefore has a theoretical effect on storage costs.We use the word "theoretical" because in the typical warehouse there is only a seventy per cent occupancy, and the saving of a few hundred cubic feet, more or less, will not materially affect overall costs. We pass this off lightly only to make the point that the real payoff to be won by efficient pallet design comes from the savings in handling.

HANDLING COSTS

The big benefit is in reducing handling costs, and this alone may be the justification for improved load design. The more units of product that can be handled at one time, the lower the unit cost. This applies equally to whatever transport is used—fork trucks, hand trucks, etc. Since the pallet is usually the basic element in handling, pallet efficiency is going to affect cost every tome that load is handled. Inefficient use of the bottom load area will increase the number of pallet loads that must be handled in order to move a given amount of product.

DEVELOPING PALLET PATTERNS

Part of the difficulty inherent in developing pallet pattern. is the fact that most companies have a wide range of package sizes. (When there are relatively few package sizes, and these fit efficiently on standard pallets, there is no problem…an unusual circumstance.) It is a very demanding job to work out a system of pallet patterns that is both efficient and easily applied by the worker on the warehouse floor. A number of pallet pattern guides are available, there are also computer programs, such as our SPACE II, which can help to generate pallet patterns. Yet it is very important to understand why and how these aids work. One weakness of all pattern books is that they fail to provide for commingled handling systems, those involving combinations of wooden pallets with some other method of palletless handling, such as clamps or slip sheets. If a pallet should be of an irregular multi-block pattern it may be impossible to pick up the load with a clamp without losing all or part of the load. Other patterns may allow clamp pickup from only one workface. If your company is blessed or cursed (depending on the viewpoint) with a large number of small packages, there are so many combinations of multi-block patterns in which the same number of packages

appear in a row of lengths as well as a row of widths, that there are as many as five different totals of packages that can be loaded onto the pallet. Without specific instructions, errors will certainly occur. This causes miscounts and losses of inventory as well as errors in unit load shipments.
For example, if a palletizer were to arrange two rows of lengths and two rows of widths, instead of the more desirable arrangement of three rows of widths and one row of lengths, he would reduce the pallet capacity by two cases per tier, or ten cases in a five high pallet (see Figure 2). It is easy to imagine the ensuing confusion at physical inventory time.

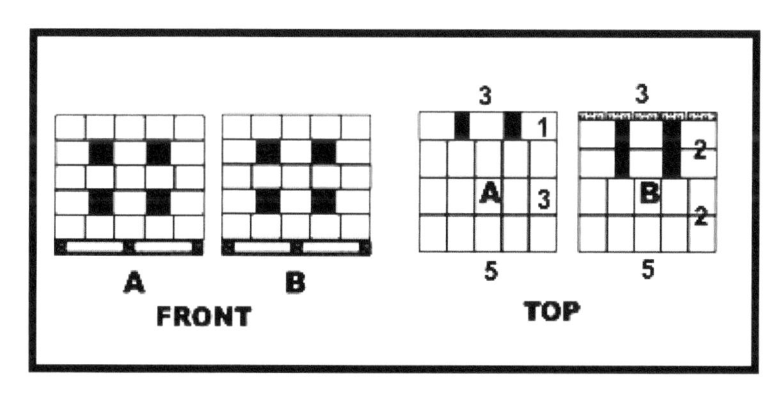

FIGURE 2

THE LOGIC OF PATTERNS

Because the best pallet pattern guide in the world cannot take into account every situation in every warehouse, an understanding of basic pattern logic will make it a lot easier to adapt the most appropriate system to individual and unique requirements. The logic of pallet patterns is based on the four all-important materials handling principles universal to all systems and usually achieving the lowest overall handling costs.

1. Handle as much material at one time as is practical.
2. Handle the material as few times as possible.
3. Use mechanical handling equipment whenever justified.
4. Develop a unit load system to achieve these objectives, and integrate that system **with** the marketing program.

Practical considerations, such as volume, frequency, distance and handling environment, are the basic determinants of which mechanical handling equipment or combinations of equipment may be used. When sufficient volume has been reached to justify a unit load program for handling and shipping, it becomes necessary to establish unit load criteria. Density and/or package dimension determine the cubic limitations of the unit load standard.

Once the maximum weight and cube measurements have been established, the objective is to use the cubic capacity of that load at maximum efficiency. Although the pallet is the most common load base in use today, it is necessary to establish criteria for unit loads that use other load bases. Whether the method is pallet, container, slip sheet, slave pallet or clamp, load patterns must still be developed.

Non-pallet handling does not relieve us of the need to provide unit load criteria. Loads may be handled with lift trucks, load arms, grabs, clamps, “Basiloid” attachments or vacuum lifters, but the objective is still to move as much each time as is practical.

SQUARE PALLET

Although loads most often have a rectangular shape (as the 48" x 40" GMA pallet), there are many instances where square or almost square loads must be handled. As an example four 55 gallon drums fit on a 48" x 40" pallet and produce a square load. Bear in mind that for pallet pattern investigations, any cylindrical object on end has only two dimensions, diameter and height. A square pallet is a good point of departure for a discussion of pallet pattern logic. Even though the load is square, it is convenient to describe load dimensions in terms of width, depth and height. The width of the load is the workface, and it is from this side of the load that a lift truck must approach to pick it up. The width of the load affects the width of the doorways and aisles through which it must pass, and vice versa. Depth determines the load center, a prime factor in determining the lift truck capacity required. Height determines how many loads may be
stacked or racked on top of one another without exceeding the available working headroom in a specific warehouse or vehicle. In spite of the fact that the above information is common knowledge in most distribution departments, many companies still plan and invest large sums
in new facilities without once considering these basic factors.

UNIFORM BLOCK

In Figure 1 (next page) the left-hand pair of patterns illustrates the most elementary unit: a uniform or regular block pattern. It is called the 'uniblock' pattern. Whether it is enclosed in a shipping container, represents a number of shipping containers, or is one of a number of uniform units stored in a vehicle, the terms and functions are the same. It is an arrangement in which all container widths are placed on the load area in one direction, and all lengths in the other. Uniblock arrangements get their tie (load stability) by alternating the direction of lengths and widths on successive tiers.

In order to demonstrate the logic inherent in pallet patterns, we will uses an example the 48" x 48" pallet with a two inch overhang on each of the four sides. This gives a load base of 52" x 52". Forgetting height for the moment, and assuming that the packages have no dimensions smaller than 5" and none larger than 52", there will by forty-five practical uniblock arrangements on the load base.

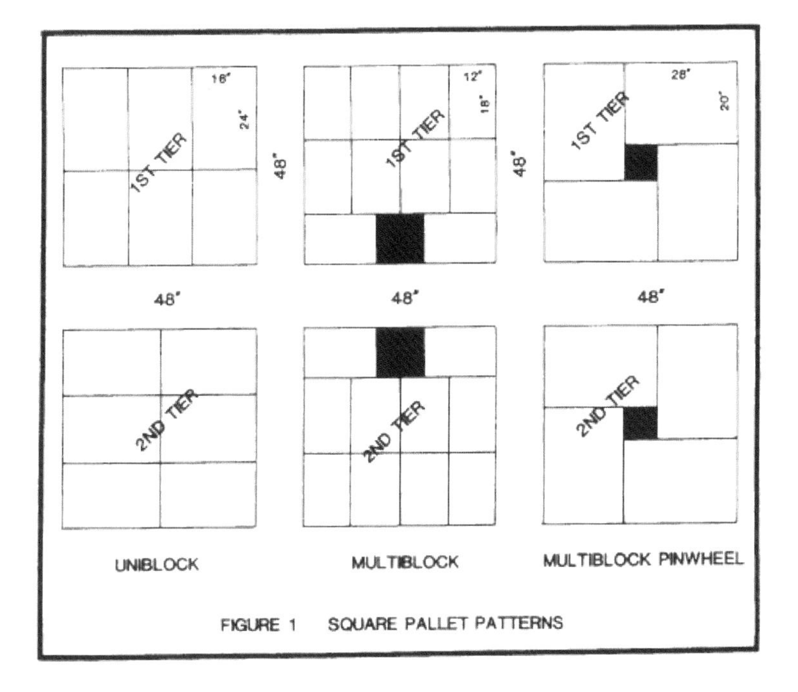

FIGURE 1 SQUARE PALLET PATTERNS

FAVORABLE DIMENSIONS

Assuming that the smallest possible carton is at least 5" in length or width, a division of the 52" base by that dimension tells us that we will never have more than ten packages on a row on either the length or width of the pallet. In practice, this is really nine, since the two end packages would overhang by almost half their lengths or widths—an unstable condition. Common sense tells us that we will come closest to 100% efficiency when we have packages with dimensions that divide into 52" without a remainder. These are called "favorable dimensions" and are found by dividing 52" by the whole numbers from one through ten. They represent all the possibilities of rows or stacks that can be arranged. Since the most basic pattern of all is a single package 52" x 52", or one per tier, we will call this condition UNITY. Two rows or stacks represent 1/2 of unity, three represent 1/3 and so on through ten. The favorable dimensions are developed in the table below.

Relation To Unity	Division	Favorable Dimension
Unity = 1	52/1	52"
One half	52/2	26"
One third	52/3	17.33"
One fourth	52/4	13"
One fifth	52/5	10.40"
One sixth	52/6	8.66"
One seventh	52/7	7.42"
One eighth	52/8	6.50"
One ninth	52/9	5.77"
One tenth	52/10	5.20"

WIDTH										
52.00"	1x1									
26.00"	1x2	2x2								
17.33"	1x3	2x3	3x3							
13.00"	1x4	2x4	3x4	4x4						
10.40"	1x5	2x5	3x5	4x5	5x5					
8.66"	1x6	2x6	3x6	4x6	5x6	6x6				
7.43"	1x7	2x7	3x7	4x7	5x7	6x7	7x7			
6.50"	1x8	2x8	3x8	4x8	5x8	6x8	7x8	8x8		
5.78"	1x9	2x9	3x9	4x9	5x9	6x9	7x9	8x9	9x9	
5.20	1x10	2x10	3x10	4x10	5x10	6x10	7x10	8x10	9x10	10x10
	52.00"	26.00"	17.33"	13.00"	10.40"	8.66"	7.43"	6.50"	5.78"	5.20"

LENGTH

FIGURE 3

STACK									
1									
2	4								
3	6	9							
4	8	12	16						
5	10	15	20	25					
6	12	18	24	30	36				
7	14	21	28	35	42	49			
8	16	24	32	40	48	56	64		
9	18	27	36	45	54	63	72	81	
10	20	30	40	50	60	70	80	90	100
52.00"	26.00"	17.33"	13.00"	10.40"	8.66"	7.43"	6.50"	5.78"	5.20"

ROW

FIGURE 4

If we put the results of this division into chart form (see Figure 3), we will get a simple table that, under ideal conditions, indicates the proper patterns for uniblock construction. This generates fifty-five possibilities, of which only forty-five are practical; as mentioned earlier, the 5.2" line should not be considered. We can also see that any package dimension that is not very close to the favorable dimensions above will produce less than desirable bottom area efficiency. With a 52" constraint and a package that falls between two of the listed dimensions, we will have to use the next larger dimension. For example, a package with an 8.5" dimension will produce exactly the same pattern as one with the favorable dimension of 8.66". In this instance the total loss does not amount to much—about 1.25" reduction overhang and a corresponding reduction in efficiency. On the other hand, a package with a 27" dimension would be reckoned as a 52" package, and this would reduce bottom area efficiency by almost 50%. In instances like this there is room for judgment; the loss of efficiency must be weighed against possible product damage through excessive overhang. One alternative is to find a more efficient pattern of another type (see Figure 1), such as a multi-block or a pinwheel. This discussion of the uniblock pattern is important as an illustration of the basic logic of pallet patterns and provides an indication of when the uniblock pattern may be used.

A LOOKUP TABLE

Multiplication of the pattern data in Figure 3 yields a useful chart (see Figure 4) which can be used as a handy lookup table to indicate to the worker exactly how many cases will be on a tier. As seen in Figure 4, the intersection of the two dimensions indicates the maximum number of packages that may be placed on a tier in the uniblock pattern. When the dimensions used are not the ones listed, the next larger set is used.

ARRANGING UNIBLOCK PATTERNS

In arranging a uniblock pattern, we always start by laying the lengths across the back of the pallet. Referring back to Figure 3, notice that length is always expressed first, then width. Length always refers to the greater dimension.

OTHER PATTERNS

By trial and error, or by the application of simple arithmetic, it's possible to work out alternative arrangements of the containers on the loading area. The multi-block pattern (see Figure 1) is nothing more than a combination of two uniblock arrangements. As we keep working with our calculator with the packages themselves, we can develop a balanced multi-block pinwheel pattern. This also consists of a combination of uniblock patterns and, although more confusing, is sometimes the only pattern that produces an acceptable utilization of the bottom area.

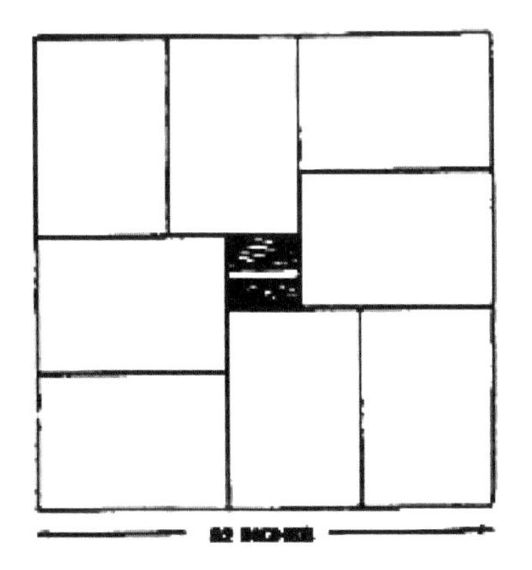

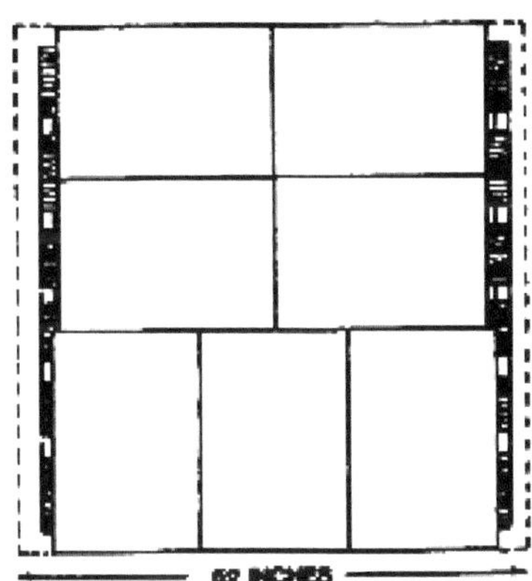

VISUALIZING THE PATTERN

One of the difficulties in establishing a proper pallet pattern by trial and error on the work floor is the difficulty of visualizing desirable overhang. Consider the example of a container 22" x 15". The worker might place the cases so as to use a full 52" in one direction, repeating the pattern until the pallet appears full but has not, in fact, been used to capacity. He has placed seven cases, a total of 2,310 square inches (22" x 15" x 7) in an available area of 2,704 square inches, leaving unused 394 square inches. Since each case uses only 330 square inches, it is to be hoped that a different pattern will allow an eighth case to be used. A pinwheel solves the problem, leaving only 64 square inches unused. This highlights the importance of using arithmetic, as well as trial and error, as an aide in the establishment of a pallet pattern; it functions both as an indicator of error and as a confirmation of correctness. In the instance above, underhang occurred on two sides. If the underhang appears on all four sides of the area, it may be even more difficult to visualize. The fact is that a two inch underhang on all four sides produces an unused area of 400 square inches, also enough space for that eighth case. Keep firmly in mind that we are talking about acceptable

load area, not pallet size, and that the man working on the load will not see a
four inch underhang but a two inch underhang. When there is a two inch underhang of the acceptable load area, he will see a completely filled pallet. The fact is that it is difficult to design
pallet patterns on the floor under conditions of work pressure; it is up to management to provide the worker with all the help he needs.

PATTERN CODES

The illustration below is a multiblock pattern in which the containers are represented by bold lines in order to illustrate a relatively simple numbering system that can describe this and many other pallet patterns. The code description is always written by beginning at the top of the pattern sketch and moving clockwise. This is analogous to starting at the rear of the pallet and moving around it to the right. It is a coding system that can easily be understood by the worker.

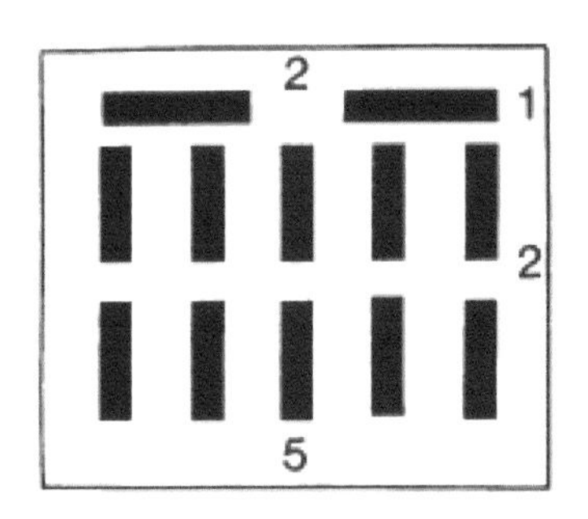

Looking at the pattern, we see a number of ways to describe it—as a combination of two uniblock patterns, or as a combination of one row of two lengths across the back of the pallet plus two rows of five widths each across the front of the pallet. In a sort of numerical shorthand, this may be further reduced to describing the pattern as (2 x 1) + (2 x 5), or two lengths times one row plus two rows times five widths. Again, note that, as a convention, length is stated first, then number of rows. The order of appearance in the code is important. If the mathematical operations are performed in the order indicated, we find that (2 x 1) + (2 x 5) = 12, an accurate

statement of the total number of containers used in the pattern. In a uniblock pattern, the number of lengths times the number of rows gives the total number of containers. If we separate a multi-block pattern into two uniblocks and tally the number in each, the sum of the uniblocks is the total number of containers in the multi-block tier. We further encode this by eliminating the mathematical symbols, so that (2 x 1) + (2 x 5) becomes 2125. If we keep in mind the format from which this was developed, we have a simple code that instructs the worker on container placement. The code also contains a ready check on the number of containers per tier. It is even more thoroughly foolproof if the code reads 2125-12 per tier. This description is always accurate and includes the correct number per tier.
Below are the codes for some of the possible patterns for a container 7" x 5" placed on a load area 48" x 36". They serve to illustrate how the use of an effective code system minimizes the job of looking up pallet patterns to find the number of containers possible on each tier.

Code #	**No. of Containers**	**Code #**	**No. of Containers**
6519	**39**	**5627**	**44**
6429	**42**	**5537**	**46**
6339	**45**	**5447**	**48***
6149	**42**	**5257**	**45**
5817	**47**	**5167**	**47**
***Best pattern**			

In each case, the simple arithmetic we have described conveys in basic form both the arrangement of the containers and the total number possible per tier. Of course, it may be that pattern 5447, which produces the greatest efficiency, may be undesirable for other reasons, such as

clamp handling or interlock. Therefore it is advisable to test any pattern before putting it into actual use. The use of the code guarantees that consistency of pattern— and therefore count—will be maintained. This alone is enough to justify the use of a pallet pattern code.
In conclusion, it must be stressed that the design of efficient pallet patterns and the discipline necessary consistently to ensure their use are proven ways of decreasing cost and increasing productivity.

A Manual Pallet Pattern Calculater

HOW TO USE THE GUIDE

Determine the length and width of the container (carton, box, etc.) and compute the bottom area. EXAMPLE:

1. Container measures 24- by 20-inches. Bottom area = 24 x 20 or 480 square inches.

2. Determine the length and width of pallet and compute the bottom area of the load.
EXAMPLE: 48 x 40 = 1920 square inches.

3. Determine the starting index (row) number.
EXAMPLE: Divide bottom area of load by bottom area of container; 1920 divided by 480 = 4. NOTE: If you want to allow for ½ inch overhang on each side of the pallet, we have prepared a special chart on the last page of this guide for finding the index number for packages up to 22-1/2 inches square (utilizing the 48- by 40-inch pallet).

4. Using the starting index number (4 in our example), locate Row 4 in the Guide. Beginning with Pattern Code 4A, scan the patterns in the row and locate the first pattern in which two container lengths are laid across the 48-inch dimension. This process will lead to the first choice: Pattern Code 4F.

5. Test the pattern for acceptable container length.
EXAMPLE: 48 divided by 24 = 2; thus, two container lengths are acceptable on the 48-inch dimension.

6. Test the pattern for acceptable container width.
EXAMPLE: 40 divided by 20 = 2; thus two container widths are acceptable on the 40-inch dimension. Note that no other pattern of four is acceptable by this simple test.

INTERLOCK

You may wish to reject Pattern Code 4F because it provides no interlock (it is a uniblock pattern). If you want to find a suitable pattern with interlock, you must move to Row 3. Pattern 3C is the obvious choice and will produce a load that measures 44- by 40-inches. You may wish to use the best pattern (4F) and tie the top of the load. If you use Pattern 3C, you will need 33% more pallets; tying the loads may be cheaper than paying for the extra pallets and extra storage space. You cannot use a pattern of five without serious overhang problems.

A METRIC EXAMPLE

In countries on the metric system, the most common pallet in use is the 120- by 100-cm pallet. Since one inch is equal to approximately 2-1/2 cm, the 120- by 100-cm pallet is the equivalent of the 48- by 40-inch pallet.

EXAMPLE: Let's assume your container measures 60 cm by 40 cm and you are using the 120- by 100-cm pallet. The bottom area of the container = 60 x 40 or 2400 sq. cm. Bottom area of the load (pallet) with no overhang is 12,000 sq. Cm. Then, 12,000 2400 = 5, which is the starting index number.

Turn to Row 5 in the Guide and scan the row for the first pattern that shows two lengths across the 120 cm dimension. We quickly can determine that Pattern Code 5F meets our requirements for length since 2 x 60 cm = 120 cm. Testing for width, we find that 40 cm is acceptable because 3 x 40 cm= 120 cm.

The final test is for the combination of lengths and widths laid across the 100 cm dimension. Our pattern meets this test since 60 cm + 40 cm = 100 cm. which is the maximum width allowed on the 120 cm by 100 cm pallet. Pattern 5F is a multi-block pattern.

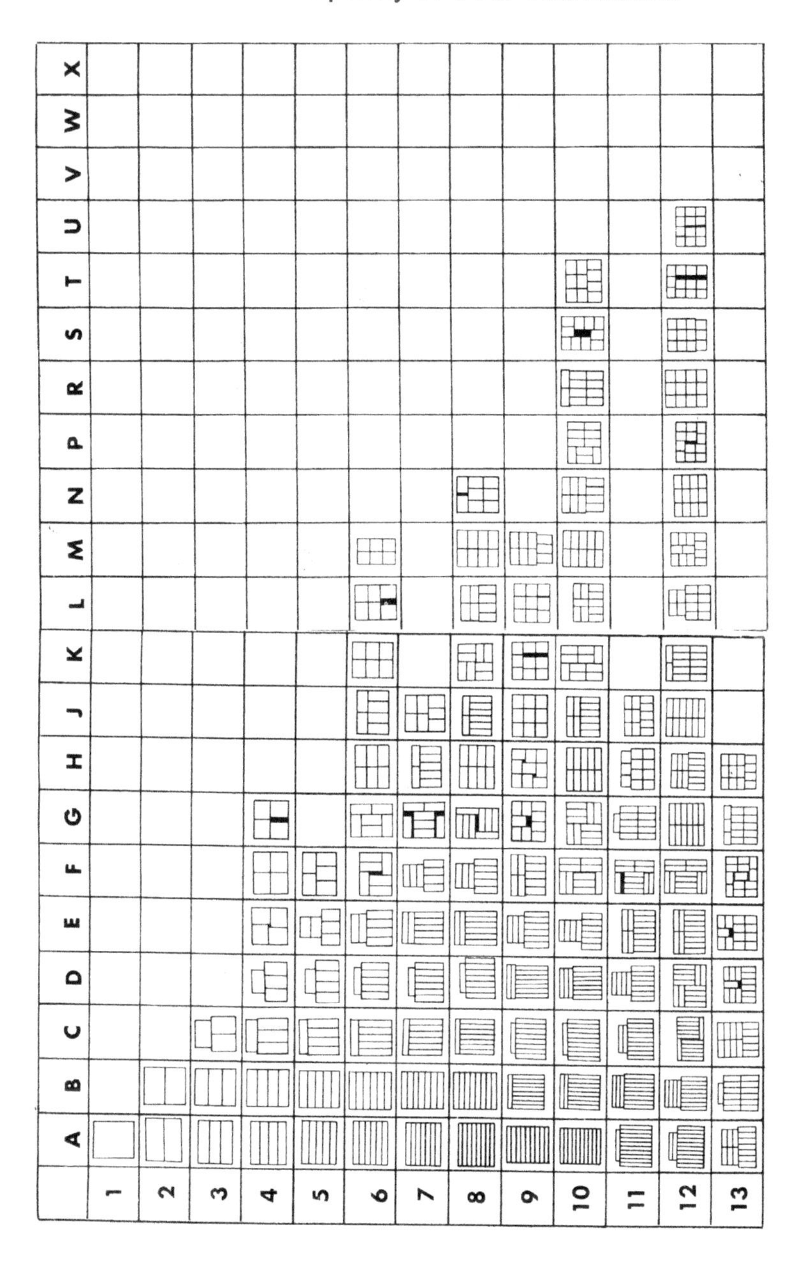
A
B
C
D
E
F
G
H
J
K
L
M
N
P
R
S
T
U
V
W
X
1
2
3
4
5
6
7
8
9
10
11
12
13

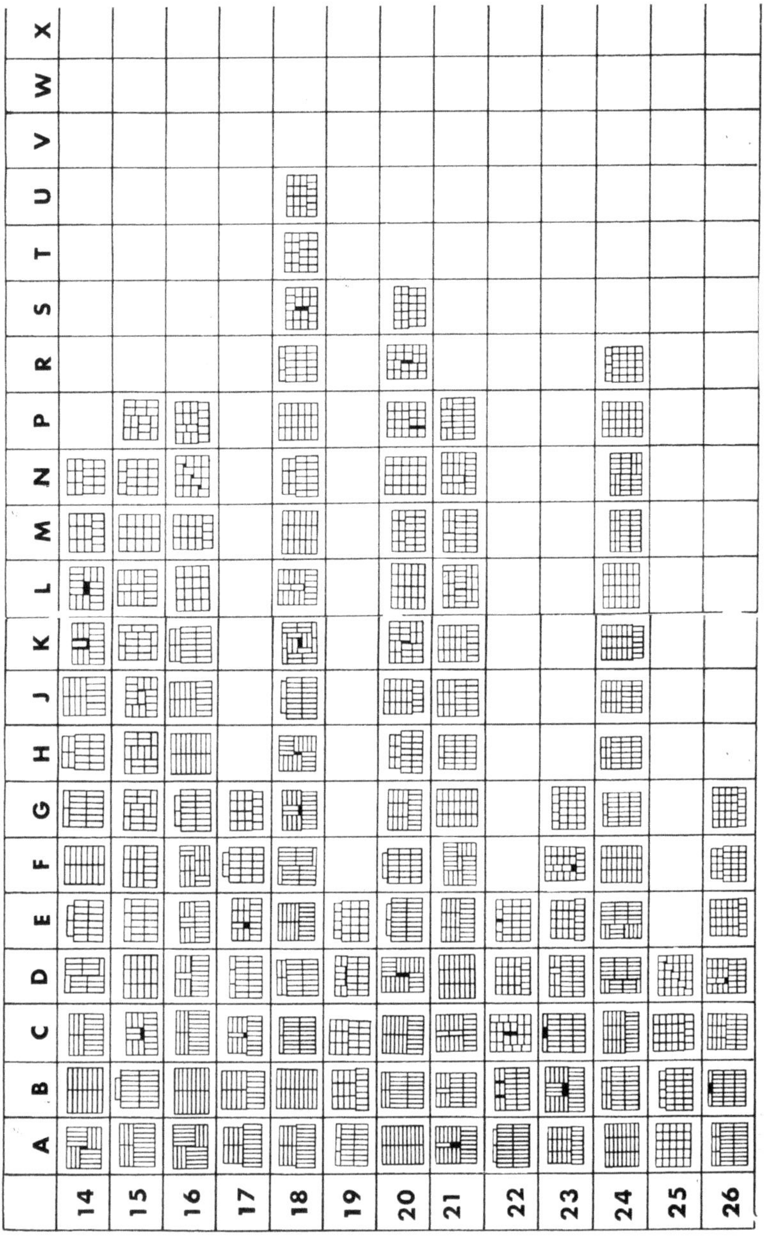

	A	B	C	D	E	F	G	H	J	K	L	M	N	P	R	S	T	U	V	W	X
14																					
15																					
16																					
17																					
18																					
19																					
20																					
21																					
22																					
23																					
24																					
25																					
26																					

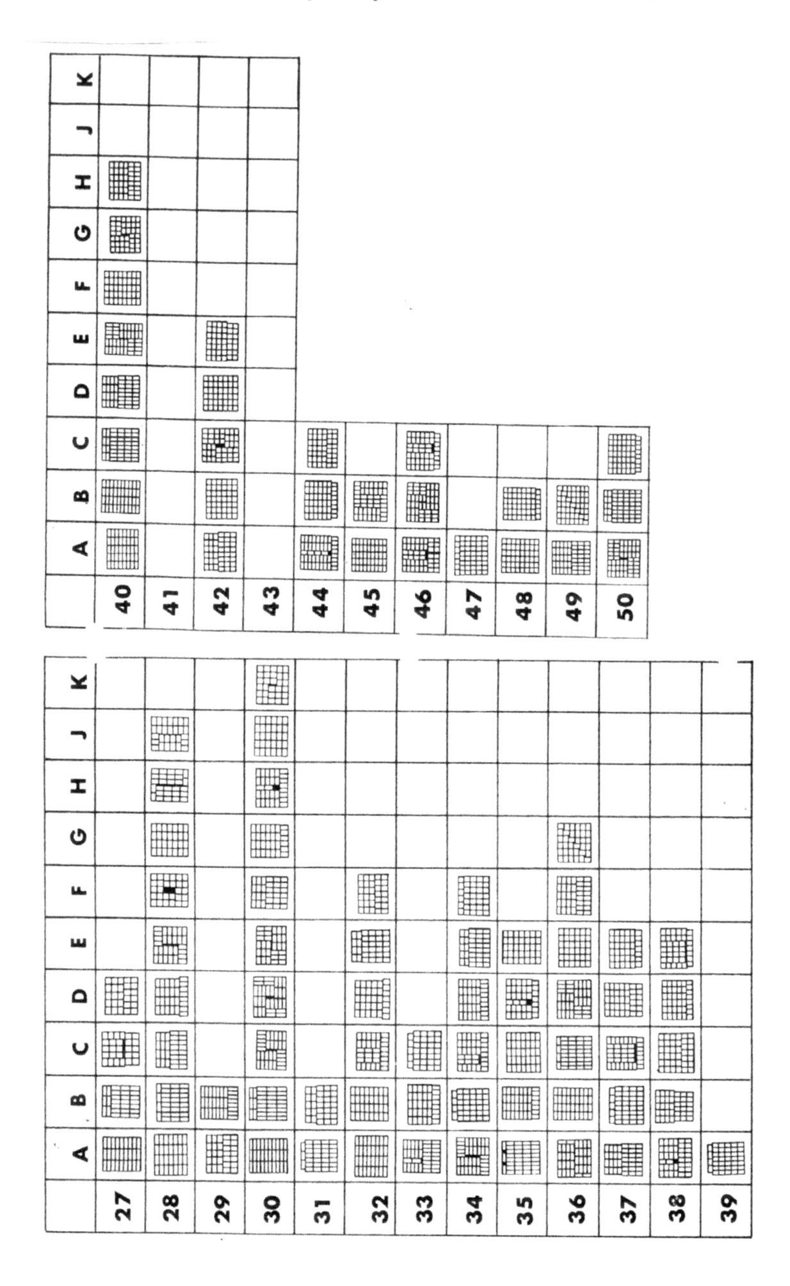
A
B
C
D
E
F
G
H
J
K
40
41
42
43
44
45
46
47
48
49
50
A
B
C
D
E
F
G
H
J
K
27
28
29
30
31
32
33
34
35
36
37
38
39

A MORE COMPLEX PROBLEM

Assume that a carton is 18 inches long and 12 inches wide. Further, assume that you will allow up to two inches of overhang on any pallet dimension. Loading area, using the standard 48 by 40- inch GMA pallet, now becomes 50- by 42 inches.

The bottom area of the load thus can be as much as 2100 sq. inches. Since the bottom area of the carton is 216 sq. inches (18 x 12), 2100 216 = 9.7, and our starting index number is 9. Turn to Row 9 in the Guide.

The first candidate pattern, 9F, must be rejected because 5 x 12 = 60 (10 inches too long), even though on the other side of the pallet 2 x 18 (36) is acceptable. Following the logic of arranging patterns according to length, let's test the next pattern, 9G. Two lengths plus one width = 48 inches; on the other side of the pallet, four widths = 48 inches. This is perfect fit with no overhang. Testing for width we find that two container widths plus one length = 42 inches. This is equal to the maximum width dimension we have allowed, so Pattern Code 9C is acceptable.

Using the same criteria we find that Code 9H is also acceptable. 9G is an irregular pattern, and 9H is a multi-chimney pinwheel pattern.

PINWHEEL PATTERNS

Pinwheel patterns are very useful, particularly for clamping loads when the sides of the containers form the chimney. Pattern 4E is an example of a square pinwheel pattern. This pattern and Pattern 8K are the only two square pinwheel patterns you will ever need when using a 48-by 40- inch pallet. Oblong pinwheels are often overlooked, and Patterns 6F, BG, 10G, and 10S are widely used. Many other pinwheel patterns are included in the Guide; some of these are rarely used because most guides do not include them. Since pinwheels are figured on the basis of combinations, you must consider the following rules when testing for oblong pinwheels:

1. The sum of the lengths and widths on each side must be less than or equal to the pallet dimensions or the allowable unit load dimensions if overhang is to be allowed.

2. When the sides form the chimney you must be certain that the sum of the widths laid across any dimension is equal to or less than the pallet dimension or the allowed load dimension including overhang.

EXAMPLE: Pattern 8G shows 6 widths laid across the longest pallet dimension. In this example, the sum of the widths should not exceed 48 inches or the load length you wish to allow. When the ends of the containers form the chimney, the opposite is the case. The sum of the lengths laid across the width of the pallet must be equal to or less than the width of the pallet.

SEE NEXT PAGE: INDEX CHART FOR DETERMINING NUMBER OF UNITS PER TIER OR LAYER

HOW TO USE THE CHART

1. **Determine the width and length of the unit or package to the nearest half inch.**

2. **Locate the length of the unit or package at the top and the width at the left side of the index chart.**

3. **The interest section of the "length" column and the "width" row shows the number of packages of this dimension that can be accommodated on the surface of 49 x 41 inches (allowing for one half-inch overhang on each side of the pallet).**

4. **In forming the unitload, the overall lateral dimensions should not exceed 49 x 41 inches.**

5. **The net height of the pallet load to be allowed (overall load height minus the height of the pallet used) divided by the height of the item to be palletized, yields the number of layers or tiers per load.**

6. **To find the best pattern shown in the pallet patterns guide, take the number of packages (intersection of "length" column and "width" row) and match it up with the corresponding row number in the guide.**

INDEX CHART FOR DETERMINING NUMBER OF UNITS PER TIER OR LAYER

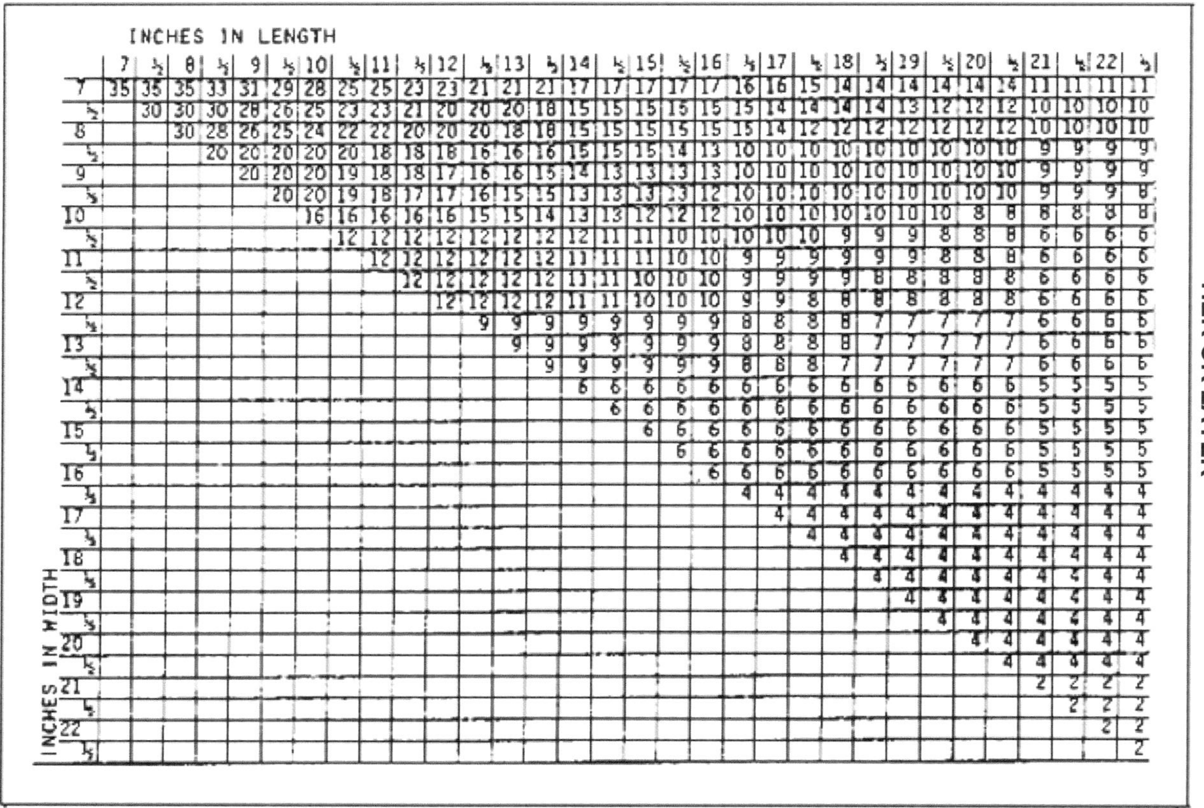

INCHES IN LENGTH

INCHES IN WIDTH	7	½	8	½	9	½	10	½	11	½	12	½	13	½	14	½	15	½	16	½	17	½	18	½	19	½	20	½	21	½	22	½
7	35	35	35	33	31	29	28	25	25	23	23	21	21	21	17	17	17	17	17	16	16	15	14	14	14	14	14	14	11	11	11	11
½		30	30	30	28	26	25	23	23	21	20	20	20	18	15	15	15	15	15	15	14	14	14	14	13	12	12	12	10	10	10	10
8			30	28	26	25	24	22	22	20	20	20	18	18	15	15	15	15	15	15	14	12	12	12	12	12	12	12	10	10	10	10
½				20	20	20	20	20	18	18	18	16	16	16	15	15	15	14	13	10	10	10	10	10	10	10	10	10	9	9	9	9
9					20	20	20	19	18	18	17	16	16	15	14	13	13	13	13	10	10	10	10	10	10	10	10	10	9	9	9	9
½						20	20	19	18	17	17	16	15	15	13	13	13	13	12	10	10	10	10	10	10	10	10	10	9	9	9	8
10							16	16	16	16	16	15	15	14	13	13	12	12	12	10	10	10	10	10	10	10	8	8	8	8	8	8
½								12	12	12	12	12	12	12	12	11	11	10	10	10	10	10	9	9	9	8	8	8	6	6	6	6
11									12	12	12	12	12	12	11	11	11	10	10	9	9	9	9	9	9	8	8	8	6	6	6	6
½										12	12	12	12	12	11	11	10	10	10	9	9	9	9	8	8	8	8	8	6	6	6	6
12											12	12	12	12	11	11	10	10	10	9	9	8	8	8	8	8	8	8	6	6	6	6
½												9	9	9	9	9	9	9	9	8	8	8	8	7	7	7	7	7	6	6	6	6
13													9	9	9	9	9	9	9	8	8	8	8	7	7	7	7	7	6	6	6	6
½														9	9	9	9	9	9	8	8	8	7	7	7	7	7	7	6	6	6	6
14															6	6	6	6	6	6	6	6	6	6	6	6	6	6	5	5	5	5
½																6	6	6	6	6	6	6	6	6	6	6	6	6	5	5	5	5
15																	6	6	6	6	6	6	6	6	6	6	6	6	5	5	5	5
½																		6	6	6	6	6	6	6	6	6	6	6	5	5	5	5
16																			6	6	6	6	6	6	6	6	6	6	5	5	5	5
½																				4	4	4	4	4	4	4	4	4	4	4	4	4
17																					4	4	4	4	4	4	4	4	4	4	4	4
½																						4	4	4	4	4	4	4	4	4	4	4
18																							4	4	4	4	4	4	4	4	4	4
½																								4	4	4	4	4	4	4	4	4
19																									4	4	4	4	4	4	4	4
½																										4	4	4	4	4	4	4
20																											4	4	4	4	4	4
½																												4	4	4	4	4
21																													2	2	2	2
½																														2	2	2
22																															2	2
½																																2

SPACE II COMPUTER PROGRAM

Many years ago, Howard Way and Associates developed a system of logic to optimize the selection of Space II. It was called Space I and was way ahead of its time. Some time later an associate transferred the logic to microcomputers. Part of the deal was that Howard Way and Associates would have lifetime use of the new program. Well, they developed an easy to use and valuable tool to optimize storage space, reduce handling, reduce the capital investment in pallets and even improve truck load-ing and transportation costs. We will outline the operational details and show you a few ways that this powerful little program can help you. If you use Windows, the use and installation will be almost trivial but we will err on the side of caution and walk you through all the details of operation. WHAT IS SPACE II? The pallet pattern calculator **Space II** is a program that uses mathematical and logical principles to interactively help to find the optimal arrangement of cases on a pallet. The basic goal is to maximize the use of the pallet area as well as to maximize the number of layers on the pallet. The original approach entailed punch cards and a look up listing that correlated with a printed pattern set. The routine took a week and cost upwards of $1000 each time it was run. Nevertheless, it was worth every cent it cost. In this Windows version, the operator can interact with the program. This allows "on the spot" changes to see the effect of different pallet sizes, case orientation and weight constraints. Now, that the program is in Windows, it is much less costly and the graphics are great!

What Can It Do? "Space II" is able to generate all of the possible patterns for stacking cases on a pallet of any size. It lists the various options, indicates the number of cases per layer, the number of layers high, the total number of cases on the pallet, the total weight, the percent utilization of the pallet area, and the per cent of utilization of the cube enclosed by the pallet dimensions . If the op-erator knows the cost of a pallet position, the program will even calculate the cost of storage per case for each solution! The program highlights the options that indicate the greatest efficiency and can print out a picture showing the true pattern to approximate scale Along each side of the sketch is a list of the lengths and widths for. that side. Given this diagram, any warehouse worker can quickly and easily duplicate the pattern with real cases. When setting up a run for a particular pallet size and case size, it is possible to place limits, or constraints, on the total weight of the pallet as well as on its total height. This makes it easy to fit present conditions or to decide on future configurations of rack opening height or load beam weight rating. Furthermore, when a pattern is affected by weight, an asterisk appears next to the listing, so that the allowable weight may be changed in order to utilize the cube more completely. The actual length and width of the pallet are also constraints, as the program will not allow overhang unless the pallet dimensions are changed to account for it. For example, a 48" x 40" pallet would be entered as 50" x 42" to allow for a total two-inch overhang of the 48" x 40" pallet.

Optimizing

An actual example is shown as (See next page), **Figure 1, How To Improve Cases Carried Per Pallet**. have used a very simple carton size of 12 inches long x 8 inches wide by 8 inches high, weighing 10 pounds, placed on a 48 inch by 40 inch pallet with no overhang . The constraints set up are that the maximum load cannot exceed 2500 pounds and the load height not counting the pallet itself may not exceed 56 inches. We entered an arbitrary description as, "printed business forms" and a cost per pallet storage location of $38.75. All of the various input data are entered in the "setup" screen shown in **Figure 1, Next Page.**

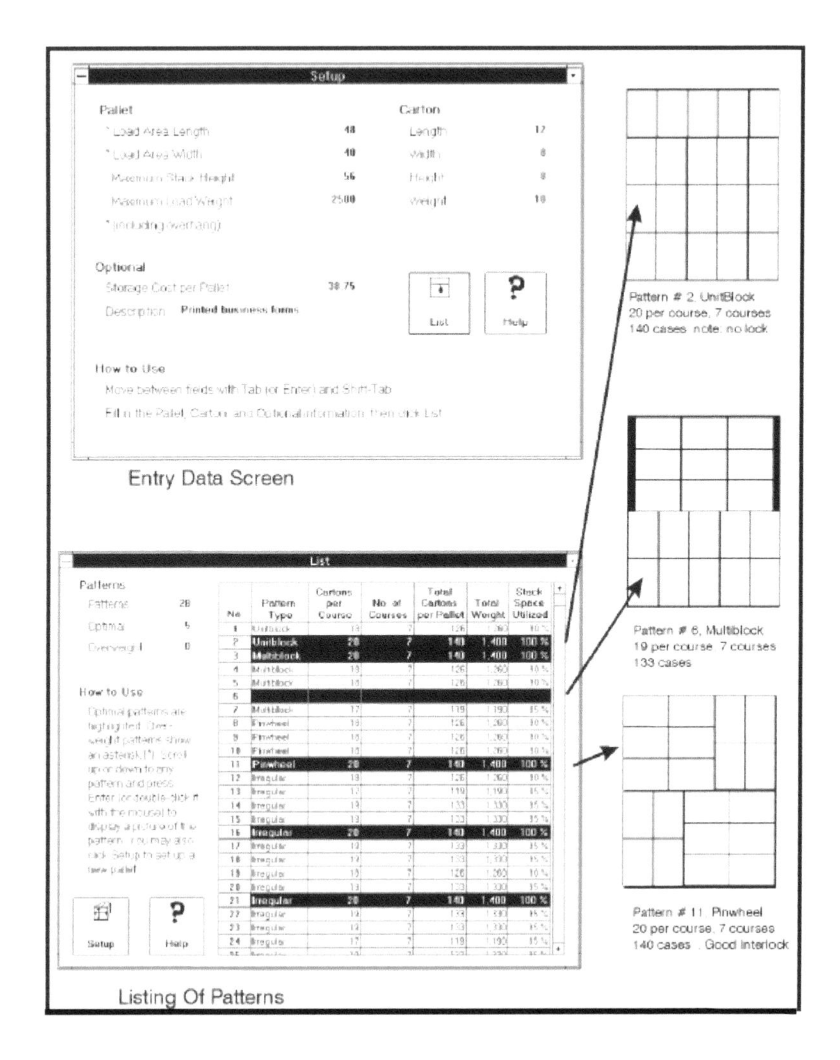

Figure 1 How to improve Cartons per Pallet

The results as shown on the "list" in **Figure 1** show that there are 28 possible patterns of which 5 appear as optimal. Weight was not a limiting factor so there are "0" overweight patterns. If weight was the limiting factor, an asterisk would show next to the pattern listing.

Which Is Best

There were five patterns that could be considered optimal. The lightly highlighted line in the “list” portion of Figure 1 shows merely where the cursor was resting. In an actual screen, that would show as a different color. Of the three pattern sketches shown , a simple logical analysis helps to choose the best. Obviously, Number 6, The multiblock solution only has 19 cartons per layer for a total of 133 cartons per pallet. Either of the other two shown (Number 11 or Number 2) are preferred from the standpoint of efficiency. Both have 20 cartons per course and 7 courses for a total of 140 cartons per pallet. However, pattern #2, “Unitblock” has no interlock to help hold the pallet together. Interlock is when each course is reversed from the one before and the cartons interlock. For this reason alone, Pattern Number 11, “Pinwheel” is the best. Pinwheel and irregular patterns are rarely developed with a ruler and a piece of graph paper. They need the computer approach! Please realize that the less obvious irregular patterns may very well hold the most cartons. If an irregular or a pinwheel is your best solution, you can give your palletizing crews graphic printouts that show exactly how to build the pallets. The use of such diagrams is highly recommended even with simple patterns. The reason of course is the achievement of consistent patterns and counts. This will reduce errors in shipments, make inventory taking easier and generally aid accuracy, space utilization and productivity.

The Display Screen

If pattern number 11 is “double clicked”, a scale diagram is generated **(See Figure 2, Scale Pattern Diagram, next page)** This pattern is a “pinwheel” and to make it even more understandable, the widths and lengths that are on each side are displayed. Detailed information about the pallet and the carton are also part of this screen on the left side.

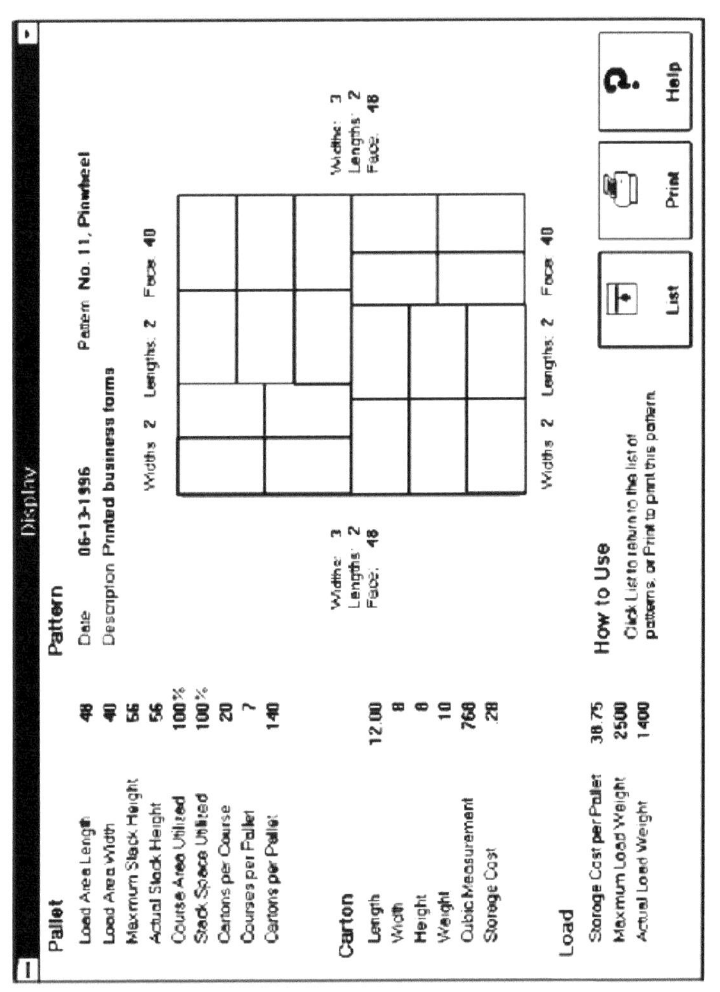

Figure 2 Scale Pattern Diagram Screen

Space Efficiency

A calculated "course" area efficiency is displayed and this indicates the relative use of the pallet surface. The Stack usage (height) is also calculated and displayed and is an indicator of desirable changes in allowable pallet height or weight. For example if the course area is not well utilized, perhaps overhang or a different pallet size may help. If the stack height is under-utilized, it may be useful to move the rack load beams a small amount to allow another course on each pallet. In either measure, the optimum pallet goal may need a change in orientation or pallet size to achieve the optimum utilization.

Cost Per Case

If the pallet storage cost is entered on the setup screen, a cost per carton or case will be displayed. This shows the bottom line advantages of pallet pattern improvement. **"the optimum pallet may need a change in orientation or pallet size to achieve the optimum utilization."**

Printout

If the mouse is "clicked" on "PRINT", a permanent scaled printout **(See Figure 3, Printout Of Optimum Pattern)** will be obtained. The quality of this print depends of course on the system printer used for windows. A complete set of laser printouts would be a great addition to any receiving/shipping department.

Variations

One of the advantages of this software over our original mainframe version is the ability to interact and try variations on a theme to come closer to the right solution. **Figure 4 "Variation Pallet A" and the display in Figure 5** illustrate a search for improvement in a different way. All of the variations involve a 48" x 40" pallet and carton dimensions of : 10.75 inches long, 6.5 inches wide, and 8.5 inches high.

The best pattern as seen in **Figure 5** is a pinwheel with 26 cases per course and 6 courses high for a total of 156 cases per pallet.

If we turn the carton on the pallet so that the length becomes 8.5 inches, the width, 6.5 inches and the height becomes 10.75, we have a whole new ballgame (Assuming that reorienting "UP" does no harm).

Now as seen in **Figures 6 and Figure 7 (Variation B),** the new best pattern has 32 cartons per course and 5 courses for a total of 160 cartons per pallet. This is a 3% improvement at no cost. More dramatic improvements are common but this is the the best procedure to find them.

Pallet Pattern Report

Pallet:

Load Area Length:	48
Load Area Width:	40
Maximum Stack Height:	56
Actual Stack Height:	56
Course Area Utilized:	100%
Stack Space Utilized:	100%
Cartons Per Course:	20
Courses Per Pallet:	7
Cartons Per Pallet:	140

Carton:

Length:	12
Width:	8
Height:	8
Weight:	10
Cubic Measurement:	768
Storage Cost:	.27

Load:

Storage Cost Per Pallet:	38.75
Maximum Load Weight:	2500
Actual Load Weight:	1400

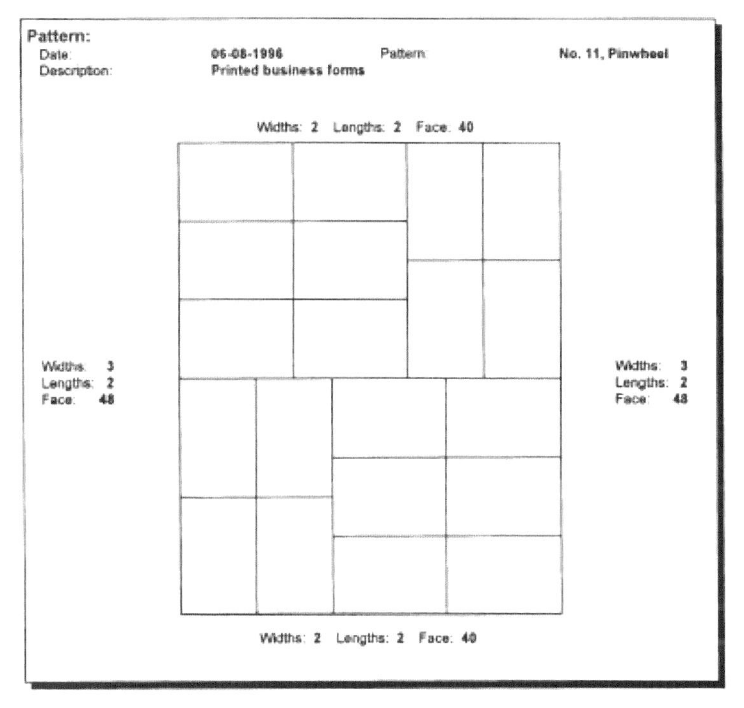

Figure 3 Printout Of Optimum Pattern

Pallet Size Variations

If there are no external restrictions on pallet size (customers, shipping, etc.), it is possible for net space gains with a larger or smaller pallet platform There is no standard pallet so think broadly. We have shown a series of trials as seen in figures 8 through 13. The pallets are 48 x40, 48 x 48, and a really huge 60 x 54. The fact is that the larger the cartons are with respect to the pallet platform, the greater the advantage to be had by a larger pallet. Yes, a larger pallet takes more space ***(See areas per load illustrations, where area per load is the footprint' of half the aisle plus the pallet, divided by the number of pallets stacked in the footprint)*** but if the gain is enough, the space per carton stored will decrease. In our example, there is an improvement going from 48" x 40" to 48" x 48". The huge pallet did not show an improvement but if the cartons were really large, the huge pallet may often prove advantageous. The comparison is seen in Figure 14, Space Efficiency Of Various Pallet Sizes. The Pallet from Figure 11 (48 x 40) is the best by almost 15.5%.

Conclusion

We like this software. Call or email us at the address below, A demo with one dimension fixed is available at no charge. The program is at a reasonable cost for a single user license. Quantity discounts are available.

Please note that the program accepts either metric or inch input.

Howard Way and Associates
PO Box 5387 , Baltimore MD 21209 USA
Phone 410-542-4446 Email: Art@Howardway.com

List

Patterns

Patterns: 37
Optimal: 3
Overweight: 0

How to Use

Optimal patterns are highlighted. Over- weight patterns show an asterisk (*). Scroll up or down to any pattern and press Enter (or double-click it with the mouse) to display a picture of the pattern. You may also click Setup to set up a new pallet.

Setup | Help

No.	Pattern Type	Cartons per Course	No. of Courses	Total Cartons per Pallet	Total Weight	Stack Space Utilized
1	Unitblock	21	6	126	1,260	69 %
2	Unitblock	24	6	144	1,440	79 %
3	Multiblock	25	6	150	1,500	82 %
4	Multiblock	22	6	132	1,320	72 %
5	Multiblock	23	6	138	1,380	76 %
6	Multiblock	24	6	144	1,440	79 %
7	Multiblock	24	6	144	1,440	79 %
8	Multiblock	21	6	126	1,260	69 %
9	Pinwheel	24	6	144	1,440	79 %
10	Pinwheel	22	6	132	1,320	72 %
11	Pinwheel	24	6	144	1,440	79 %
12	**Pinwheel**	**26**	**6**	**156**	**1,560**	**86 %**
13	Irregular	23	6	138	1,380	76 %
14	Irregular	23	6	138	1,380	76 %
15	Irregular	22	6	132	1,320	72 %
16	Irregular	22	6	132	1,320	72 %
17	Irregular	25	6	150	1,500	82 %
18	**Irregular**	**26**	**6**	**156**	**1,560**	**86 %**
19	Irregular	24	6	144	1,440	79 %
20	Irregular	24	6	144	1,440	79 %
21	Irregular	24	6	144	1,440	79 %
22	Irregular	22	6	132	1,320	72 %
23	Irregular	24	6	144	1,440	79 %

FIGURE 4.List for "Variation Pallet A"

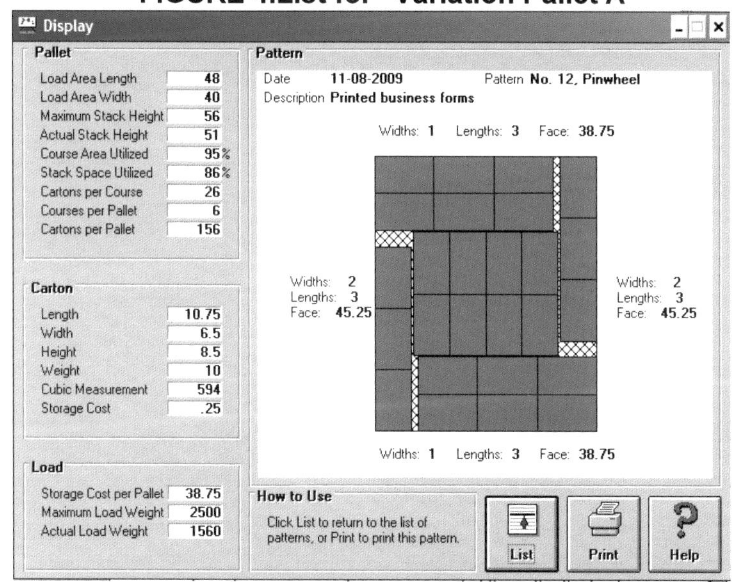

FIGURE 5. Display for "Variation Pallet A"

List

Patterns

Patterns 58
Optimal 8
Overweight 0

How to Use

Optimal patterns are highlighted. Over- weight patterns show an asterisk (*). Scroll up or down to any pattern and press Enter (or double-click it with the mouse) to display a picture of the pattern. You may also click Setup to set up a new pallet.

Setup Help

No.	Pattern Type	Cartons per Course	No. of Courses	Total Cartons per Pallet	Total Weight	Stack Space Utilized
1	Unitblock	28	5	140	1,400	77 %
2	Unitblock	30	5	150	1,500	82 %
3	Multiblock	31	5	155	1,550	85 %
4	Multiblock	29	5	145	1,450	80 %
5	Multiblock	27	5	135	1,350	74 %
6	**Multiblock**	**32**	**5**	**160**	**1,600**	**88 %**
7	Multiblock	30	5	150	1,500	82 %
8	Multiblock	28	5	140	1,400	77 %
9	Multiblock	30	5	150	1,500	82 %
10	Pinwheel	28	5	140	1,400	77 %
11	Pinwheel	29	5	145	1,450	80 %
12	**Pinwheel**	**32**	**5**	**160**	**1,600**	**88 %**
13	Pinwheel	30	5	150	1,500	82 %
14	Pinwheel	28	5	140	1,400	77 %
15	Pinwheel	30	5	150	1,500	82 %
16	**Pinwheel**	**32**	**5**	**160**	**1,600**	**88 %**
17	**Pinwheel**	**32**	**5**	**160**	**1,600**	**88 %**
18	Irregular	27	5	135	1,350	74 %
19	Irregular	28	5	140	1,400	77 %
20	Irregular	29	5	145	1,450	80 %
21	Irregular	28	5	140	1,400	77 %
22	Irregular	29	5	145	1,450	80 %
23	Irregular	30	5	150	1,500	82 %

FIGURE 6. List for "Variation Pallet B"

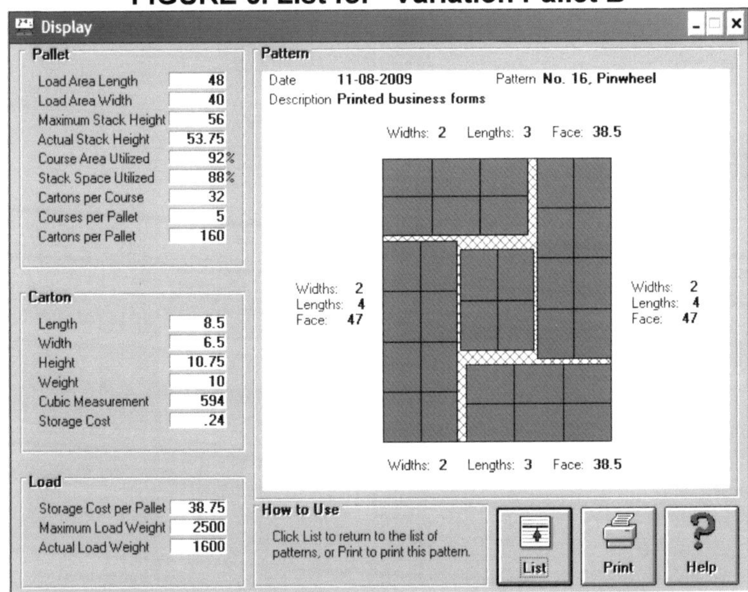

FIGURE 7. Display for "Variation Pallet B"

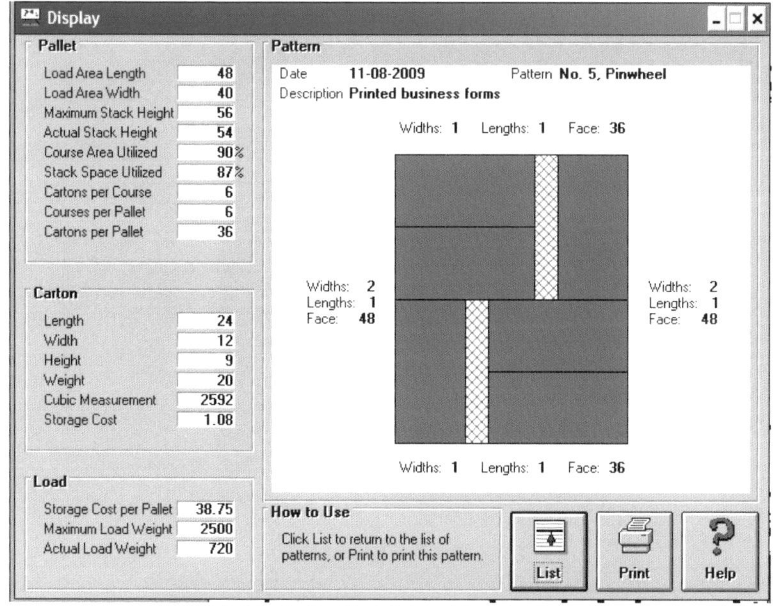

FIGURE 8. Pallet Size Test 48 x 40

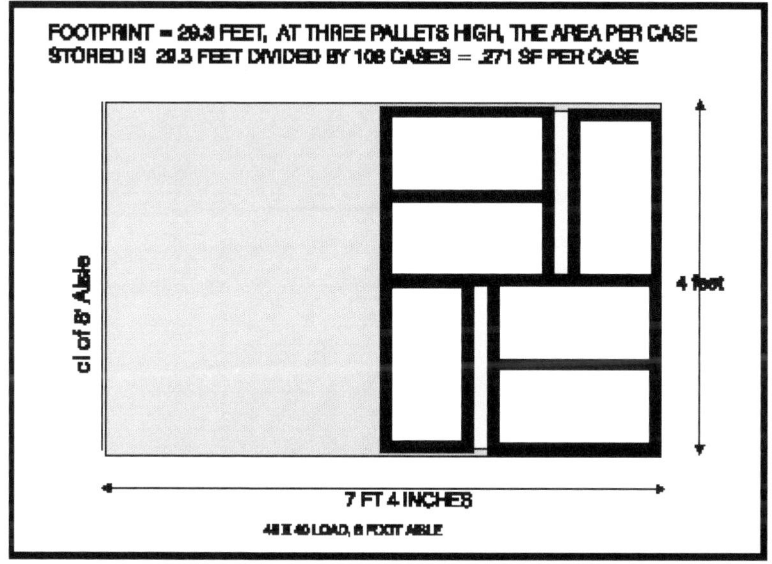

FIGURE 9. Area Per Load 48 x 40

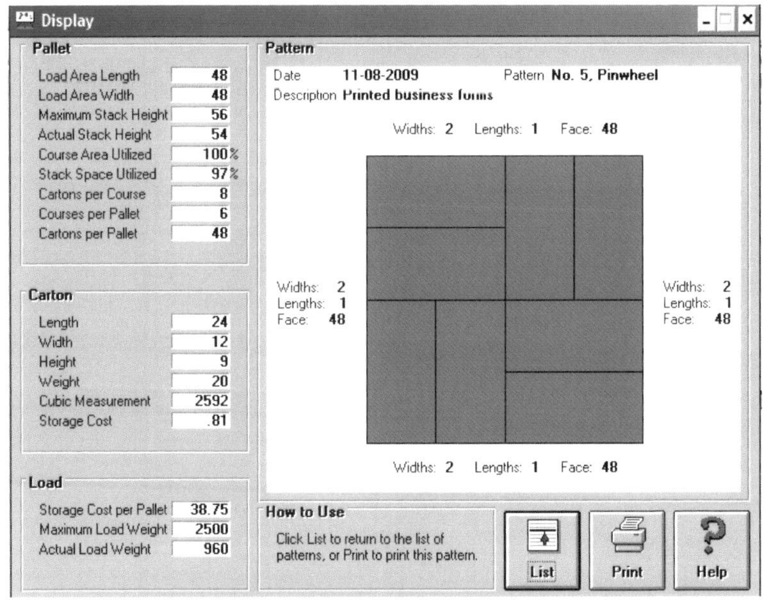

FIGURE 10. Pallet Size Test 48 x 48

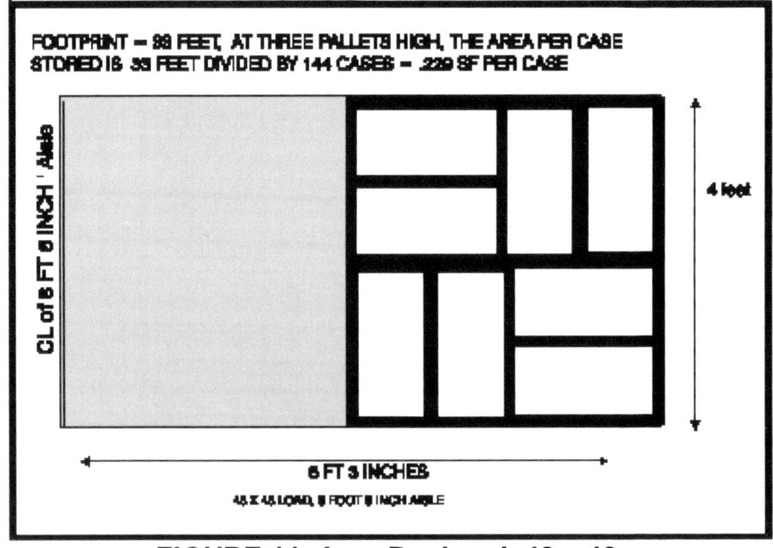

FIGURE 11. Area Per Load 48 x 48

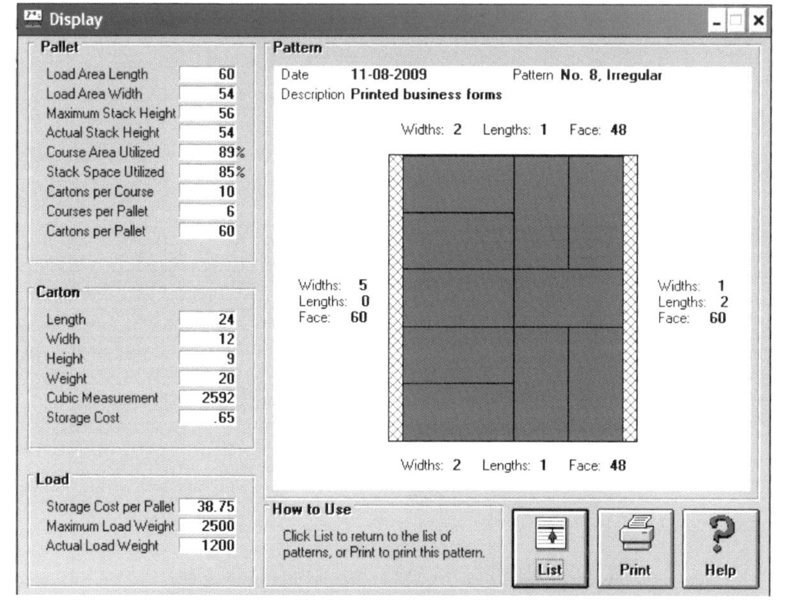

FIGURE 12. Pallet Size Test 60 x 54

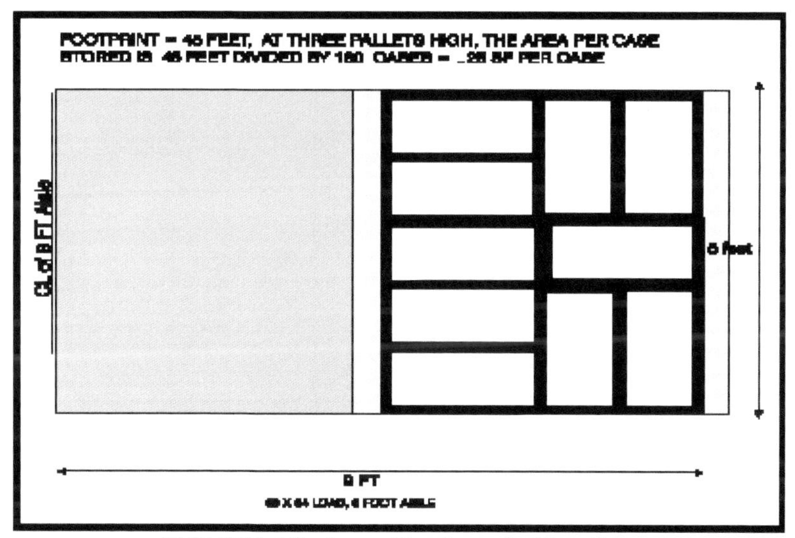

FIGURE 13. Area Per Load 60 x 54

FIGURE 14. Space Efficiency Of Various Sizes

PALLET SIZE RESULTS					
FIGURE NUMBER	PALLET SIZE	AISLE WIDTH	CASES PER PALLET	CASES IN 3 HIGH	AREA PER CASE
FIGURE 9	48 x 40	8 FT	36	108	.271SF
FIGURE 11	48 x 48	8FT 6 IN	48	144	.229SF
FIGURE 13	60 x 54	9 FT	60	180	.250 SF

Questions, Comments, and Consulting Arrangements:

Howard Way and Associates
PO Box 5387 , Baltimore MD 21209 USA

Phone 410-542-4446
Email: Art@Howardway.com

CPSIA information can be obtained
at www.ICGtesting.com
230680LV00001B

* 9 7 8 0 9 8 0 0 7 3 4 5 4 *